SOLAR ENERGY IN THE 80s

Other Pergamon Titles of Interest

ANDRESEN, A. F. & MAELAND, A. J.	Hydrides for Energy Storage
ARIZONA STATE UNIVERSITY LIBRARY	Solar Energy Index
DE WINTER, F.	Sun: Mankind's Future Source of Energy
DIXON, A. E. & LESLIE, J. D.	Solar Energy Conversion
EGGERS-LURA, A.	Solar Energy for Domestic Heating and Cooling
EGGERS-LURA, A.	Solar Energy in Developing Countries
FAZZOLARE, R. A. & SMITH, C. B.	Changing Energy Use Futures (4 Volumes)
GARDEL, A. A.	Énergie: Économie et Prospective
GOODMAN, L. J. & LOVE, R. N.	Biomass Energy Projects
HOWELL, D.	Your Solar Energy Home
HUNT, S. E.	Fission, Fusion and the Energy Crisis, 2nd Edition
MCVEIGH, J. C.	Sun Power, 2nd Edition
MESSEL, H.	Energy for Survival
O'CALLAGHAN, P. W.	Building for Energy Conservation
O'CALLAGHAN, P. W.	Energy for Industry
OHTA, T.	Solar-Hydrogen Energy Systems
ROSS, D.	Energy from the Waves, 2nd Edition
SECRETARIAT FOR FUTURES STUDIES	Solar Versus Nuclear: Choosing Energy Futures
STARR, C.	Current Issues in Energy
VEZIROGLU, T. N.	Solar Energy: International Progress
VEZIROGLU, T. N. & SEIFRITZ, W.	Hydrogen Energy System (5 volumes)
VEZIROGLU, T. N. et al.	Hydrogen Energy Progress (4 volumes)

Pergamon Related Journals (Free specimen copy gladly sent on request)

Energy, the International Journal
Energy Conversion and Management
International Journal of Hydrogen Energy
Journal of Heat Recovery Systems
Solar Energy
Space Solar Power Review
Sun at Work in Britain
Sun World

SOLAR ENERGY IN THE 80s

Proceedings of the Conference on Solar Energy
held in London, 14-15 January 1980

Edited by

COSTIS STAMBOLIS

Organised by

HELIOTECHNIC ASSOCIATES INTERNATIONAL, LONDON

PERGAMON PRESS

OXFORD · NEW YORK · TORONTO · SYDNEY · PARIS · FRANKFURT

U.K.	Pergamon Press Ltd., Headington Hill Hall, Oxford OX3 0BW, England
U.S.A.	Pergamon Press Inc., Maxwell House, Fairview Park, Elmsford, New York 10523, U.S.A.
CANADA	Pergamon of Canada, Suite 104, 150 Consumers Road, Willowdale, Ontario M2J 1P9, Canada
AUSTRALIA	Pergamon Press (Aust.) Pty. Ltd., P.O. Box 544, Potts Point, N.S.W. 2011, Australia
FRANCE	Pergamon Press SARL, 24 rue des Ecoles, 75240 Paris, Cedex 05, France
FEDERAL REPUBLIC OF GERMANY	Pergamon Press GmbH, 6242 Kronberg-Taunus, Hammerweg 6, Federal Republic of Germany

First edition 1981

British Library Cataloguing in Publication Data

Solar energy in the 1980's.
1. Solar energy - Congresses
I. Stambolis, Costis
II. Heliotechnic Associates International
621.47 TJ810 80-41617

ISBN 0 08 026123 X

In order to make this volume available as economically and as rapidly as possible the author's typescript has been reproduced in its original form. This method has its typographical limitations but it is hoped that they in no way distract the reader.

Printed in Great Britain by A. Wheaton & Co. Ltd., Exeter

CONTENTS

Introduction — vii
C STAMBOLIS

Past, present and future developments in solar energy- a personal view — 1
H TABOR

Energy from biomass- solar energy through biology — 5
D HALL

Solar prospects for the UK and the EEC — 21
J. K. Page

The role of solar energy and energy conservation — 23
R STEWART, B MURPHY and J HEALEY

The low energy adaptable dwelling (lead) project — 63
D FORREST

Every man's solar energy — 69
A RUSSELL-COWAN

Considerations for establishing the durability of solar absorber surface coatings — 79
M G HUTCHINS

State police barracks at Bagnols-Sur-Ceze with solar hot water and solar space heating systems — 93
G CHOULEUR

An overview of work carried out by solar energy developments — 107
D MICHAELIS

Establishing a community of solar heated homes — 115
R TUCKER

Solar architecture in Colorado on the threshold of the 80's — 123
P J TABB

Solar energy applications in Greece — 133
D DASKALAKIS

Some solar energy projects in India — 139
S M JAUHRI

Solar passive houses in Ladakh (N. India) — 153
C STAMBOLIS, S WATSON and S JAUHRI

Solar powered systems: Short review of solar powered facilities and components developed by Dornier system — 157
K SPEIDEL

Solar power generation by solar ponds — 189
H TABOR

Photovoltaic conversion of solar energy in the 80's 197
Y CHEVALIER

The department of energy's solar energy research and development programme 203
E G BEVAN

US international commercialization programs for solar energy 215
R C SPONGBERG

UNESCO's solar energy activities in developing countries 219
T BERESOVSKI

Power to the people 227
B HOLDSWORTH

Solar energy utilization and its implications for trade unions 233
C LEVINSON

Index of Contributors 235

INTRODUCTION

As our social and economic infrastructures tend to become closely related to the planet's energy equation the need for the development and application, on a large scale, of peaceful and renewable energy sources is becoming more apparent. In a world which is threatened with the non-availability of fuel oil supplies and the menace of nuclear war, one is bound to look into the future energy prospects with considerable scepticism. On the other hand, the role of renewable energy sources, such as solar energy, has not yet been fully appreciated, nor explored, by most of the world's governments and major industrial organisations.

Taking a global energy view, we feel that solar energy is one of the most promising alternative energy sources because of its diffuse nature and worldwide availability. It is true that some countries enjoy greater amounts of solar radiation than others and this is why different types of systems and components are required in each case. Unfortunately, solar energy is not yet seen, by the majority of nations, as a national energy source. There is still a lot of mystique attached to the words 'solar energy', let alone its applications. It is also true that during the passed decade interest and funding on solar energy increased manyfold, and that phenomenal progress was achieved in a considerably short time. However, a lot more needs to be done if solar energy is to make a world impact as a major energy source.

Looking ahead into the 80s and beyond, we see that the major challenges will be the development of economic solar technologies. Until the time comes when economic interpretation of life becomes more human, the problem for solar energy will always be the competition with conventional energy sources. But already there are many places in the world today where cost-effectiveness is irrelevant simply because no other energy sources exist.

This volume contains papers that were presented in the 'Solar Energy in the 80s' conference which took place in London at the very beginning of the new decade. The great majority of the work reported at the conference was based on many years practical experience with the result that the ensuing discussions concentrated on the actual current and potential problems of solar energy applications. It may seem unconventional for a non-academic group such as Heliotechnic Associates International to have convened a conference of this kind but it was precisely because of the practical orientation of and focusing on working installations that H.A.I. took the initiative to organise this event. The conference which received wide support and publicity was sponsored by four diverse organisations, a fact clearly indicating the type of support that solar energy activities have at long last started to receive. The four sponsoring organisations were: UNESCO, one of the oldest UN bodies which started supporting solar energy activities more than 25 years ago; the Atmospheric Sciences Research Center of the State University of New York at Albany, a leading US research centre on solar energy; the Williams & Glyn's Bank Ltd, a major British financial institution; and the Atelier, an established engineering consulting organisation in St. Albans, Hertfordshire.

In many respects the conference proved to be an eye-opener, as indeed most conferences are supposed to be, with a fair amount of work reported here for the first time. Harry Tabor, one of the pioneers in the field of solar energy, announced an important breakthrough in solar power generation: the use of a non-convecting solar pond to operate a 150 kW thermal power station which makes use of a low temperature, organic-vapour, Rankine turbine as the power plant. (See pages 189-196). The design, installation and operation of such a pilot plant will certainly stand as a milestone in the history or harnessing the sun, heralding the development of much larger units in the 5MW range, which can find application in many parts of the world, especially in developing countries. On the electricity generation side the rapid progress and immense potential prospects of photovoltaic cells was reported by Yves Chevalier, at COMES, the French official Government Agency on Solar Energy (see pp 197-202). With a total of 2 MW worldwide installed photovoltaic capacity at present and a predicted 100 MW by 1985 and prices dropping in real value, solar cells may well prove to be one of the most economic ways of producing electricity in many remote areas and used for equipment with low power requirements.

Several new solar buildings were presented and perhaps the most exciting one was the police barracks at Bagnol-sur-Ceze in the south of France (see pp 93-106) Architect George Chouleur has used the entire south facing vertical wall of the building to integrate 155 m^2 of collector surface. Thus the quite awkesome problem of accommodating the collectors has been overcome. Solar architects Sham Jauhri, Dominic Michaelis and Philip Tabb described many new solar housing schemes, most of them having been realised over the last year. In the papers presented by Ron Stewart (see pp 23-62) and Don Forrest (see pp 63-68) several successful ways were shown of combining and integrating solar energy components to overall energy conservation schemes in residential and industrial buildings. Such a novel approach could well pave the way for energy savings on a big scale in urban areas where many restrictions exist for the location of solar collector areas. Klaus Speidel of Dornier System, announced for the first time the development of a new solar powered water pump (see pp 157-188) and solar cooling and desalination projects which have taken place in the Middle East. Such developments highlighted the tremendous potential for applying solar energy systems in developing countries which usually enjoy high annual solar radiation values but often lack the necessary economic and industrial resources.

Professor David Hall discussed the importance of biomass in a global context and critically evaluated the different methods which are currently used for producing useful fuel such as methane, ethanol and alcohols (see pp 5-20). He also stressed the fact that the total world wood-fuel consumption is grossly underestimated in officially published statistics and could be as large as three times more than is usually acknowledged. It is surprising that approximately ½ of the world's total energy use is biomass. However, there are imminent dangers from the excessive use of wood in many countries which are afflicted from desertification. It is now estimated that half of all the trees that are cut down are used for cooking or heating. Perhaps one of the most practical ways of utilising solar energy is through the photosynthetic process and although efficiencies are usually very small, the potential for increased wood-fuel production is enormous because of the great versatility of plant growth.

The social and economic implications of solar energy utilisation and the impact of growing solar markets were expertly debated by Dr. Charles Levinson (see pp 233-234). In his presentation Dr. Levinson emphasised the role of controls in matters of energy policy and how these are held by the large multi-national energy corporations and financial institutions. Therefore it is not surprising that research into softer technologies, particularly of solar energy, has been stunted and diverted towards the development of centralised collection and generating systems. The democratic power generation opportunities which are now presented by solar energy utilisation are likely to be threatened by efforts to dominate and monopolise this field by a few large companies.

The decision making aspects in the field of solar energy are perhaps far more crucial than most people who are involved in the broad energy field can appreciate. With new processes, materials and techniques being announced almost every day the need is becoming apparent for a critical evaluation of this rapidly evolving technology. For decisions to be taken as to the type of equipment or process that are worth developing, the type and size of investment required, and where to market and locate manufacturing facilities, require a sound understanding of the long term implications. In view of the fast increasing demand for the use of solar systems there is a need for authoritative information on systems' performance and an examination as of the various choices and constraints which are inherent in solar energy utilisation. A range of these choices are analysed in the various papers contained in this publication. Although the technological choices are wide it must be pointed out that, at least at this stage, these are important economic and social constraints.

Clearly, decisions are taken by different people serving under different capacities, and the following is an outline of the main levels involved:

Consumer	- house owner, developer, farmers, local government authorities, industry, engineering and building contractors;
Industrial	- manufacturers of equipment (e.g. collectors) and producers of prime materials (e.g. copper);
Corporate	- financial corporations, banks, investment advisors;
Government	- legal, economic and policy making officials.

The decisions which usually need to be taken and where information is required are of the following kind: (a) type of system, (b) type and size of equipment, (c) level of investment, sales policy, (d) operational conditions and service, (e) legal and social aspects, (f) economic subsidies.

The papers presented at this conference which was purposely subtitled 'A decision maker's conference', outline, and some of them in great detail, the technological options that are currently available and others that are being developed. Options do exist and it is up to the 'decision makers' to evaluate the solution that they offer. A good example of the kind of decisions that will need to be taken is given by Professor John Page in his paper on 'Solar Prospects for the UK and the EEC (see pp 21-22). Prof. Page raises the question of scale, in solar systems, and the need to concentrate on medium scale applications. He says: "Small may be beautiful, but medium size is likely to prove substantially better".

It appears that the most critical period in the development of solar energy utilisation is going to be the decade of the 80s. The question immediately arises as to the directions that we should follow so as to lead solar technologies to a position where they can make substantial contribution to world energy needs. What has become clear over the last five years is that too much money has been spent in re-inventing the wheel and in improving component efficiency at the expense of economics. As Tabor rightly observed (see pp 1) "The place for sophistication is in the thinking, not in the hardware". The directions that solar technologies should then follow in their further development should be towards improving simplification, reliability and economic viability. Compared to other modern technologies solar energy has a long way to go but the foundations have already been laid. What solar enthusiasts should be arguing about is not how many percentage points solar technologies can score before the year 2000, but how fast they can become a leading force in electricity supply, domestic and industrial heating/cooling, cooking and agricultural applications for the millions of needy people in the Third World.

Looking ahead into the 80s we can confidently say:

> Use small scale solar energy now, intensify large scale solar research and have faith in the eventual use of the sun as major energy source.

conc

Costis Stambolis
London, June 1980

PAST, PRESENT AND FUTURE DEVELOPMENTS IN SOLAR ENERGY—A PERSONAL VIEW

H. Tabor

The Scientific Research Foundation - Jerusalem, Hebrew University Campus, P.O. Box 3745, Jerusalem, Israel

If Peter Glaser, whom we had expected to participate,(1) had been here, he would - if my memory serves me correctly - have been the third of the present participants who attended the First World Symposium on Solar Energy, held in Arizona in 1955, the other two being our session chairman, John Page and myself. We were, of course, younger then than we are now and naively enthusiastic. I presume to be speaking for my two colleagues when I say that we are still solar enthusiasts but our enthusiasm is somewhat tempered by a better understanding of the issues involved, some of which I shall touch on in these remarks.

We can divide the 25 years that have elapsed since the Arizona meeting into two periods: the first twenty i.e. up to 1974 and the fuel-oil crisis - and the five years since then.

In the first period, there was a small amount of activity, solid and basic: it was, in effect, an extension of the excellent work started at M.I.T. by Prof. Hottel and his colleagues in the forties. There was almost no government sponsorship in countries such as the U.S., UK, Germany, Japan, etc. and indeed most other countries. There was some governmental support in France - largely because of the group of heliotechnicians who had been working in Algeria and returned to France when Algeria became independent - and also because of the personality of Prof. Felix Trombe. There was some government support in Australia - within the CSIRO - again mainly due to the persistence of one man - Roger Morse - who was head of the Mechanical Engineering Division of CSIRO. There were government-sponsored projects in some other countries but all on a rather small scale: the Soviet Union also reported on solar R & D. In Israel, the work of my own group came directly under the prime-minister, Mr. Ben Gurion, who, if I am not mistaken, was the only prime-minister or president in the world to actively promote solar research - and this was in the 50's! In the U.S., UK and other countries, there were small university groups - with even smaller funds - such as the Wisconsin University group inspired by the late Prof. Farrington Daniels. A solar conference, in this period might bring together fifty scientists and engineers. The work reported was mostly of good scientific standard but relatively unsophisticated in the technical sense.

In the second period, all this changed: in 1975, at the Los Angeles ISES meeting, 4000 people turned up: the cynics implied that the large attendance was prompted by the possible availability of large R & D funds rather than by a compelling desire to solve the energy crisis, but be that as it may, a large number of individuals and groups rapidly became involved in solar work.

This second period has been characterised by an increasing degree of sophistication, some of which may be good but much of which is harmful. Thus new computational techniques

(1) This lecture was added to the programme when it became known that Dr. Glaser would not be able to attend.

have been developed to determine rapidly the expected yield of a solar installation - and this is good, but can go too far: the last ISES meeting in 1979 revealed a very large amount of intellectual effort being applied to refining such methods far beyond what is reasonable, seeing that the original methods were only statistical and the sunshine itself exhibits wide variations from year to year, and to making the systems more sophisticated. This may have been - at least at some stage - due to the influence of space scientists and the aero-space companies entering the solar field and it is my personal view that they missed the point.

In 1977, I had been given figures for a total installed cost of \$40 per ft^2 of collector for a house heating installation (this being incremental to the cost without solar) and whilst the figure has been contested by some, it has been confirmed by others. Yet the annual useful heat yield of the collectors was as low as 60,000 BTU/ft^2 in a monitored Canadian installation to about 200,000 BTU/ft^2 under better conditions. At \$5/million BTU - the cost of heat from oil at that time - the value of the solar yield was 30 cents per ft^2 per year in the bad case, to \$1 per ft^2 per year in the good case. The capital cost of \$40/ft^2 appears indeed as a very poor investment. What the sophisticates forgot was that we are dealing with an energy source of very low unit value i.e. under \$1 per square foot of delivered solar energy - to use the example given: to invest \$10/ft^2 for the collector (a current target price) is borderline economically, but to let this rise to \$40/ft^2 for the installed system is rendering many solar energy applications totally unviable economically, converting them to playthings in an affluent society[(2)].

The place for sophistication is in the <u>thinking</u>, not in the hardware. Thus, who is important is the bio-chemist, photo-chemist or physicist who builds a highly sophisticated model of how a photon-converter works, this possibly leading to simple hardware, with a chance for low cost. A recent example is in photo-electro-chemical cells where good science at the Weizmann Institute has raised the efficiency of conversion from 2-3% to 8%. The photobiologists, when they fully understand photosynthesis and plant growth may devise efficient converters (plants) of extremely low cost. Yet another example - that I will describe in another session of this conference - is the solar pond, which is extremely sophisticated in conception but which leads to the lowest-cost thermal solar collector yet developed.

On the whole, most solar devices are not really viable, even against the present high cost of fuel. The exceptions are almost the same as they were 25 years ago, e.g. the simple thermosyphon solar hot-water heater which has no pump, controllers or maintenance costs. (As soon as an attempt is made to upscale this simple water heater - for example to heat a house - the extra sophistication eats away the economic viability). Passive buildings - where the building is designed to exploit solar radiation for heating and - perhaps - nocturnal radiation for cooling, is another exception (provided the enthusiasts don't go too far: many passive buildings are now costing too much!) What we - to our shame - unfairly call "primitive" societies knew remarkably well how to build livable structures centuries ago: we are just having to relearn much of that primitive "technology"!

Earlier, I referred to Prof. Trombe in France. His work is widely recognised (he was awarded the Farrington Daniels Award of the International Solar Energy Society last year). His two main areas have been passive houses and solar furnaces: the latter had nothing to

---

(2) Affluent in money, not in primary energy sources!

do with the energy crisis. Trombe, a high-temperature chemist, needed a clean source of high-temperature heat, so he built solar furnaces including a large one involving a field of articulated mirrors concentrating solar radiation to a focal point on a high platform. Later, the aerospace people, possibly fascinated by the beauty and structural engineering of Trombe's furnace, proposed building such systems to generate power[(3)]. The concept will work but economic and operational viability appear - at least to me - very remote.

To look to the future, we note the two basic characteristics of solar energy that are the cause of most of the difficulties to date:

(i) The intensity is low, so that very large devices are required to collect significant amounts of energy.

(ii) Solar radiation is intermittent.

These two characteristics provide the clue as to where we have to concentrate our efforts i.e.

(a) materials technology - in the widest sense;

(b) energy storage.

Just as man in the western world insists on his private car - so as not to feel enslaved by a railway timetable - so he demands energy on tap when <u>he</u> wants it, not when the sun shines. Thus, low-cost energy storage (accent on low-cost) is an a priori condition for solar energy utilisation on any significant scale, most particularly long-term storage. This is one of the basic motivations of the very good work being done on bio-conversion as a major long-term contribution for solar energy utilisation. Bio-conversion is also a special case of materials technology - how to get large areas of "collector" for a small capital outlay. The use of phase-change materials for heat-storage is again a question of materials technology: not only must they have desirable reversible properties, but ultra low-cost materials have to be identified - especially if long-term storage is considered. Other aspects of materials technology are in the materials of construction of solar devices. Examples: new semi-conductors for low-cost direct photo-conversion. Note that here again, the problem of energy storage is an inhibiting factor to viable direct conversion: we do not have, today, an effective and cheap method of storing electricity on a large scale. A further example is new plastics, that need to be resistant to the temperatures and ultra-violet radiation associated with thermal solar collectors.

I will conclude with a philosophical comment: there are no magical solutions to the energy shortage problem: in any area the energy needs will be met - in whole or in part - by a <u>mix</u> of energy sources - and any contribution by solar - even if not ideal - is to be welcome and exploited to the full.

---

(3) A form of "central receiver" had been proposed by the Russians in the forties, but the major inspiration for current U.S. activity would appear to have been the work of Prof. Francia in Italy who demonstrated small-scale, power-tower generators in the sixties.

ENERGY FROM BIOMASS—SOLAR ENERGY THROUGH BIOLOGY

D. O. Hall

King's College, 68 Half Moon Lane, London SE24 9JF, UK

ABSTRACT

The current worldwide government sponsored research and development of solar energy in 1980 approaches $2½ billion. More than four-fifths of this is allocated to expenditures in the Americas (Brazil, USA and Canada) where important biomass programmes have been implemented over the last five years - the result of the realisation that biomass can provide liquid and gaseous fuels, besides solid fuel and chemicals, at a cost and with socio-economic benefits that look surprisingly favourable even at these early days of implementation studies and pilot plants. Biomass for energy programmes are currently being implemented and are under assessment in many countries of the world.

KEYWORDS

Biomass, energy, solar energy, fuels, resources, conversion, energy ratios, economics, national programmes.

INTRODUCTION (refs. 1-17; 21-25; 41-45)

Hardly a day goes by without there being a news item warning us of the impending shortage of oil and what it is going to cost us - if we can get it! This belated realisation that non-renewable liquid fuels are going to increase in price, and possibly even be rationed, is one of the main reasons why biomass is being looked at so seriously by so many of the developed countries. Possibly even more important, for the developing countries of the world, is the so-called 'woodfuel crisis' which may be even more serious since the problems of deforestation have such long-term detrimental agricultural, social and economic consequences.

The oil/energy problem of the last five years has had three clear effects on biomass energy use and development. Firstly, in the developing countries there has been an accelerating use of biomass as oil products have become too expensive and/or unavailable. Secondly, in a number of developed countries large research and development programmes have been instituted which have sought to establish the potential and costs of energy from biomass. Estimated current expenditure is approaching $100 million per annum in North America and Europe. The work is still at the early stages but results look far more promising than was thought even two years ago. Demonstration projects and small-scale commercialisation are being rapidly implemented. Thirdly, in at least one country, viz. Brazil (which currently spends over a half of its foreign currency on oil imports), large scale biomass energy schemes are being implemented as rapidly as possible - the current investment is over half-billion dollars per annum.

There is no doubt that the majority of the people in the world live by growing plants and processing their products. The main issue in developing countries is that of scarcity and the problem of trying to maintain, or possibly even to increase, the present level of use without harming agricultural or forestry and ecological systems. More efficient use of existing biomass and possible substitutes for biomass use, e.g., solar and wind based technology, should be considered and implemented as quickly as possible to reverse the trend of excessive biomass use, as is already occuring in many countries. In the developed world the expertise exists and is already being used to implement biomass energy programmes from the point of view of potential technology and economics. Biomass can provide a source of energy now and in the future; just how much it can contribute to the overall provision of energy will very much depend on existing local and national circumstances and thus it is imperative that each country establish its energy use patterns and the potential of biomass energy. This is not very easy to accomplish quickly, but needs to be done as soon as possible.

Not many people need reminding that our fossil carbon reserves, whether for fuel or chemicals, are all products of past photosynthesis. However, what most people do not realise is the magnitude of present photosynthesis - it produces an amount of stored energy in the form of biomass which is about ten times the world's annual use of energy. Table 1 also shows that the total amount of proven fuel reserves below the earth is only equal to the present standing biomass (mostly trees) on the earth's surface while the fossil fuel resources are probably only ten times this amount. This massive-scale capture of solar energy and conversion into a stored product occurs with only a low overall efficiency of about 0.1% on a worldwide basis but because of the adaptability of plants it takes place and can be used over most of the earth.

It is not widely appreciated that one-seventh of the world's annual fuel supplies are biomass (equivalent to 20 million barrels of oil a day - the same as the USA consumption rate) and that about half of all the trees cut down are used for cooking and heating. Because this use is mostly confined to developing countries it has until recently been sadly neglected and the effects of overuse of biomass are having serious and long term consequences. In the non-OPEC developing countries, which contain over 40% of the world's population, non-commercial fuel often comprises up to 90% of their total energy use. This non-commercial fuel includes wood, dung and agricultural waste and because of its nature is seldom thoroughly considered. Total wood-fuel consumption is probably three times that usually shown in statistics, and about half of the world's population relies mainly on wood for their cooking (four-fifths of total household energy use) and heating. Furthermore, supply statistics of non-commercial energy can be out by factors of 10 or even 100.

In my estimation a rural person in the developing countries uses on average about 15 GJ of biomass-derived energy every year. This is the equivalent to 1 tonne or 1.4 m^3 of air-dry wood. Local and regional differences in annual use abound, as do the relative proportions of wood, dung and agricultural wastes. The developing countries have a population of about 3 billion of which about 70% are rural. Biomass energy in rural areas usually supplies more than 85% of the energy - and this is mostly used in the household for cooking. There is also an urgent need to supply more local energy for agriculture and small industry. We calculate an annual worldwide rural biomass energy use of about 3.2×10^{10} GJ. In Africa about 65% of the total energy consumed is biomass-derived, in Latin America the figure is about 45%, while in India and the Far East about 50% is biomass energy. Including the urban population of developing countries, who often use large quantities of biomass-derived energy (say, an average of 8 GJ/year), and the use of biofuels by small scale industries, we come up with a total biomass energy use in developing countries of about 4×10^{10} GJ; this is about one-seventh of the world's total energy use - equivalent to about 20 million barrels of oil per day.

In this paper I would like to present some evidence that fuels produced by solar energy conversion are a very important source of energy now and will continue to be so for the foreseeable future - probably even to an increasing extent. We should re-examine and if possible, re-employ the previous systems; but, with today's increased population and standard of living, we cannot revert to old technology, but must develop new means of utilising present-day photosynthetic systems more efficiently. Solar biological systems could be realised to varying degrees over the short and long term. Some, such as the use of wood, biological and agricultural wastes, and energy farming, could be put into practice immediately, whereas others may never become practicable. Photobiological systems can be tailored to suit an individual country taking into consideration total available energy, local food and fibre production, ecological aspects, climate and land use. In all cases the total energy input (other than sunlight) into any biological system should be compared with the energy output and also with the energy consumed in the construction and operation of any other competing energy producing system.

Solar energy is a very attractive source of energy for the future but it does have disadvantages. It is diffuse and intermittent on a daily and seasonal basis, thus collection and storage costs can be high. However, plants are designed to capture diffuse radiation and store it for future use. Hence the serious thought (and money) being given to ideas of using biomass as a source of energy - especially for liquid fuels, but also for power generation and other end uses (Table 2). I am aware of biomass programmes in the UK, Ireland, France, Germany, Denmark, Sweden, USA, Canada, Mexico, Brazil, Sudan, Kenya, Australia, New Zealand, India, Philippines, Thailand, Israel, South Korea and China. The biggest difficulty with implementing them seems to be the simplicity of the idea - the solution is too simple for such a complex problem! Fortunately for us, plants are very adaptable and exist in great diversity - they could thus continue indefinitely to supply us with renewable quantities of food, fibre, fuel and chemicals. If the serious liquid fuel problem which is predicted within the next 10 to 15 years comes about, we may turn to plant products sooner than we expect. Let us be prepared!

What I am definitely not proposing is that any one country will ever be able to derive all its energy requirements from biomass - this is highly unlikely except in especially favourable circumstances. What each country (or even region) should do is to look closely at the advantages and problems with biomass energy systems - summarized in Table 3. The long term advantages are considerable but implementation of significant programmes will take time and require important economic and political commitments. The programmes will vary in their emphasis and thus most of the research and development should be done locally. Such R & D is an ideal opportunity to encourage local scientists, engineers and administrators in one field of energy supply. One should always be fully aware of the assumptions involved in any energy cost projections (especially if they are more than 6 to 12 months out of date!) before extrapolating or drawing firm conclusions. Even if biomass systems do not become significant suppliers of energy in a specific country in the future, the spin off in terms of benefits to agriculture, forestry, land use patterns and bioconversion technology are, I think, significant.

FACTORS FOR IMPLEMENTATION

The main factors which will determine whether a biomass scheme can be implemented in a given country are (a) the biomass resource, (b) the available technology and infrastructure for conversion, distribution and marketing, and (c) the political will combined with social acceptance and economic viability. These points are now considered in turn. (ref. 1)

(a) The resource base

The total annual production of biomass, (net primary production), the amount of wood produced (including natural forest and managed plantations), and the harvested weight of the major starch and sugar crops are shown in Table 4. In addition there is a worldwide availability of crop residues and other organic wastes. Although the amount of such wastes have been calculated in some detail for the USA, Canada and some European countries where they have been identified as the major short term biomass-resource, such figures are not generally available for the developing countries. Such data that is available is often questionable and cannot, at present, form a basis for any energy planning discussions. In addition to established sources of wood and food a wide range of other land and aquatic cultivation systems have been proposed for the future.

(b) Technology for conversion

Biomass as it stands in the field or collected as wastes is often an unsuitable fuel since it has a high moisture content, a low physical and energy density and is incompatible with present demands for a fuel to be used in internal combustion engines - the main power source for transport and agriculture in most countries. Established conversion technology can be divided into the biological and the thermal (Table 2). The great versatility of biomass energy systems is one of their most attractive features - there are a range of conversion technologies already available (and being improved) yielding a diversity of products, especially liquid fuels to which the world seems to be addicted and on which most world economies have recently been based.

Plant materials may be degraded biologically by anaerobic digestion processes or by fermentation, the useful products being methane, ethanol and possibly other alcohols, acids and esters. At present the established technologies are the anaerobic digestion of cellulosic wastes to form methane or the fermentation of simple sugars to form ethanol. The most suitable feedstocks for anaerobic digestion are manures, sewage, food wastes, water plants and algae.

The most suitable materials for thermal conversion are those with a low water content and high in lignocellulose, for example wood chips, straw, husks, shells of nuts, etc. The most likely processes to be adopted will use part of the material as fuel for the production of the required mixture of carbon monoxide and hydrogen (synthesis gas) for the subsequent catalytic formation of alcohols and hydrocarbons. During gasification oxygen or steam may be introduced in order to enhance the degree of conversion to synthesis gas and to increase its purity.

Two basic routes of catalytic conversion, of synthesis gas to further products can be recognised. The gas may be converted directly to hydrocarbons via the Fisher-Tropsch synthesis, or may be used for the formation of methanol. Both routes are well established in connection with use of gas produced from coal with plants operating in countries such as South Africa and Germany. Some plants using sorted domestic rubbish are operating and considerable research is being carried out on gasification of wood. On a smaller scale commercial wood-fuelled gasification plants have been available for some time, the gas produced being suitable for use in stationary engines.

Hence the technology exists for the production of heat, steam, electricity, gas, and liquid transport fuels from biomass. In addition alcohols are the conventional starting point for a wide range of low molecular weight chemicals, plastics and fibres. Implementation of such schemes depends on local requirements, economics, energy balances, etc. This great versatility of biomass conversion technologies is one of its distinct advantages.

(c) Energy ratios and economics

In an ideal world the main factors to be considered in adopting a specific biomass route would relate to the energy gain and the economics. The benefit to be derived by converting plant material to ethanol can be expressed in terms of the net energy ratio (NER) which is obtained by dividing the final yield of energy in useful products by the total energy inputs derived from sources other than the biomass itself. In computing the inputs, in addition to fuel, fertilizer and irrigation, a value has to be assigned to the farm and process machinery and to ongoing maintenance. In general a net energy gain is seen where the fermentation and distillation is powered by the burning of crop residues, as in the case of sugar cane, or by burning of wood obtained from close by - as for a cassava alcohol-distillery powered using *Eucalyptus* wood. Reported NER values for such systems vary from about 2.4 to over 7. For most starch crops and sugar beet the values are close to or below one, i.e., more energy is used than is produced. However, this may still be worthwhile if the fuel source is for instance cheap coal or poor quality, wood or residues, etc., which are in effect converted to a high quality fuel.

For the thermal conversion routes an efficiency can be calculated as the ratio of energy in the end product as a fraction of the energy content of the starting material. Since part of the feed is completely combusted to power the conversion this value must be less than one. Here the justification is again related to the production of a high quality, higher energy density liquid fuel, from a bulky wet biomass source. At present the efficiency of methanol production from wood is probably about 25%, however efficiencies of around 60% are theoretically feasible.

The estimates of the cost of producing alcohol by ferementation of biomass vary enormously from 10 US.¢ per litre to over 60¢ per litre. However, many of these estimates are based on paper studies. Realistic figures from Brazil (1979) are as follows: 30.5¢ per litre for sugar cane alcohol and 31.7¢ for cassava-derived alcohol as compared to gasoline at an ex-refinery selling price of 23¢ per litre and a retail price of 39.6¢ per litre. The alcohol prices are FOB distillery selling prices calculated for alcohol produced by autonomous distilleries computed to yield the investor a 15% annual return on investment calculated according to the discounted cash flow method and on PROALCOOL funding of 80% of the fixed investment at 17% annual interest.

Ethanol production, from farm crops, in the USA is profitable at present due to the tax structure. The Federal Government has passed an exemption of gasoline tax on GASOHOL (a 10% ethanol : gasoline blend) equivalent to $0.4 per gallon. Various states have further tax incentives so that in Iowa for instance the combined subsidies work out at over $1 per gallon. The justification for this lies in the fact that in order to maintain corn prices the government subsidizes each bushel of corn NOT produced with one dollar. A bushel of corn can produce 2.5 gallons (US) of ethanol to be used in 25 gallons of gasohol.

Most paper studies indicate that methanol produced by gasification of wood and catalytic resynthesis will be considerably cheaper than ethanol produced by fermentation. The only problem is that no production plants are operating at present. A detailed analysis for a methanol plant in New Zealand can be summarized as follows. At an efficiency of 50% for a 2500 oven-dry tonnes per day plant at 1977 prices using N.Z. national cost benefit economics (10% on capital, DCF over 30 years, no tax or depreciation) the product price was $214 per tonne using wood at $55 a tonne or $146 a tonne for wood at $25 a tonne. These values are equivalent to product costs of between 17 and 19¢ a litre, comparable with those summarized recently by the U.S. Solar Energy Research Institute (SERI) where methanol costs from wastes or fuel crops varied from 11¢ to 35¢ per litre at raw material costs from a negative value for waste to about $50 a tonne, with assumed efficiencies of methanol production of between 25 and 50%.

(d) Implementation

The assessment and implementation of biomass energy programmes in individual countries is an excellent opportunity for a country to develop its own research, development and demonstration capabilities in this area. The types of biomass available for conversion to energy are very much region-dependent, e.g., sugar cane and cassava in hotter climates, cellulose in temperate areas and hydrocarbon shrubs in arid zones. No one country has a monopoly on biomass-for-energy expertise and indeed it is widespread - note the ethanol programme in Brazil, the biogas plants in China and India, gasifiers in Germany, straw burners in Denmark, agro-forestry in East Africa, village woodlots in Korea and parts of India, and so on. It is also an opportunity to develop collaboration between scientists, engineers, foresters, agronomists, sociologists, economists, and administrators within a country, within regions and between countries. Biomass conversion to energy is an "old-but-new" and rapidly developing area which interests many young scientists, engineers, etc., because it has both immediate practical and also longer-term basic research and development requirements. It would be worthwhile recognising and encouraging local talents and imagination and putting them to use as soon as possible in field, laboratory and pilot-scale projects.

It is imperative that good energy assessments be made in individual countries (with or without outside help), identifying the energy flows, limitations in data, and opportunities. The priority needs must be matched with the available resources. The practical importance of biomass energy must be made very clear. Proposals to implement such biomass energy systems must be clearly spelt out. Such data must then be used to influence the decision makers, only after the socio-economic and institutional barriers are clearly recognised and proposals are available on how they can be overcome and with the all important, full co-operation of the people themselves and the decision makers - otherwise experience shows that nothing may be accomplished.

CURRENT STATUS OF BIOMASS PROJECTS THROUGHOUT THE WORLD

Brazil (ref. 35,41)

By far the most ambitious biomass programme which has been planned is that in Brazil for the production of alcohol from sugar cane, sorghum and cassava. This National Alcohol Programme (PNA or Proalcool) was established in November 1975. The alcohol will be used to blend with petrol - mixture of up to 20% (by volume) requires no adjustment to the engine (over the past 10 years the State of Sao Paulo with over 1.3 million cars has varied the alcohol content of its petrol up to 18% - depending on the availability of alcohol and price of molasses). In 1977, 141 new alcohol distilleries were authorised by Proalcool which would require an investment of about $900m and would supply 3.2 x 10^9 litres of alcohol by 1980, about a fifth of the projected gasoline requirement. By 1985 total production of alcohol could reach 10 x 10^9 litres with a huge total investment of billions of dollars. An economic analysis of the production of alcohol from sugar cane and cassava calculated selling prices as fuel, ex-distillery, of $0.33/litre, 81% of the present retail price of gasoline on a volume basis. Brazil is clearly embarking on an ambitious programme of fuel import substitution using the natural advantages of land and climate which it has - and it may be a very useful demonstration to other countries and provide exportable technology and expertise. The indirect benefits such as saving foreign exchange, creating new employment, encouraging domestic technology and industry, and reducing pollution, are great.

U.S.A. (refs. 13-20,27)

The USA has a very large R, D & D programme on biomass which had a budget of $56m for 1979 and will be even greater for 1980. The 1975 US use of energy was 71 x 10^{18}J; the total standing forest inventory however has an energy content three times this annual energy use. The biomass which could be used to produce energy and is being looked at very carefully includes forests, specifically grown crops and plantations, and agricultural, industrial and urban wastes of all types.

Flexible plant systems are also being considered - thus biomass intermediates (besides food) are processed into fuels or materials depending on relative price levels. Two examples cited are corn and sugar cane production. It is 'theoretically possible to reorganise the present US corn biomass system to permit the production of 10 - 18 x 10^9 l of ethanol (or its equivalent in other fermentation products) while obtaining the same quantity of end-use food products in the form of beef, poultry and pork'. The ethanol could be blended at 10% with a ¼ of the US gasoline. 'Gasohol' is currently widely available at over 2,000 gasoline stations and is priced cheaper than gasoline. Sugar cane produces numerous byproducts such as molasses, alcohol and bagasse which can be used for power generation, fermentable substrates and substitute wood.

The concept of intensive silviculture biomass farms or energy farms has been one subject of detailed analysis. Fast growing deciduous trees which coppice (resprout from stumps) have been examined, as have numerous other trees including those which fix nitrogen to ammonia.

In Table 6 we have summarised two (of many existing) studies of biomass products and their costs to give an idea how close some technologies are to economic reality. The assumptions implicit in all these calculations must, however, be recognised and carefully analyzed.

Canada (refs. 28,29 and 41)

Canadian studies on the large scale production of methanol from biomass show that by 2025 up to 42% of the transport fuels could be provided by such means. Methanol represents a rather unique fuel combining the portability of liquid petroleum products and the clean, even-burning characteristics of natural gas. The Federal Government recently announced a $180m programme on biomass R & D and also provided economic incentives to industry to use wood and forest residues to provide energy.

Europe (Refs. 33,34,41,42,50)

In Europe a number of countries and the EEC are conducting extensive feasibility studies of the potential which biomass may have for supplying a source of energy and fuels in the future. Trial plantings of alder, willows, poplars, etc., are being undertaken in addition to assessing energy yields from agricultural residues, urban wastes, techniques of conversion, waste land and forest potentials and algal systems. Biological and thermal conversion equipment is available and is in great demand for use in Europe and for implementing overseas programmes.

Sahel region (Refs. 10 and 24)

The problem of deforestation and desertification has highlighted the lack of fuelwood in most countries of this region. It is a scarcity which is reaching crisis proportions in large parts of South Asia, the Sahel, Andean countries, Central America and the Caribbean.

The per capita requirement for cooking alone is about 0.5 m^3 of wood/year. Total fuelwood requirement is probably about 1 ton which is equivalent to about 400 kg per person per year - equivalent of a 3 inch lump of coal per person per day. Demand from urban areas and industry to use charcoal instead of the more expensive kerosene and oil is diverting woodfuel from rural areas, resulting in increasing forest and scrub destruction (often long distances from villages) and use of dung as a fuel instead of as a fertiliser. A recent Dutch study of the Sahel region points out two possible solutions: decreasing fuelwood demand by using stoves which reduce consumption by 70%, and increasing the supply of fuelwood by establishing 'forest plantations'. Costs of reforestation and fuelwood production have been calculated for the tree species suggested and the conclusion is that 'under the conditions assumed it is an economically feasible activity'. Naturally there are institutional problems which impinge on agriculture and other practices of the society but if such countries are to achieve even a modicum of internal fuel production they should seriously consider such biomass systems and set up fuelwood plantations as soon as possible.

India (Refs. 12 and 37)

India has a long-standing biogas programme and is rapidly implementing new approaches to biomass utilization. A study in Tamil Nadu, South India, on the possibility of growing Casuarina (a nitrogen fixing tree) as a source of fuel for a power plant generating 100 MW electricity has shown favourable economic and social benefits. In direct competition with coal a payback period for such an energy plantation would be from 15 to 30 years. An equivalent of 110 hectares is required to generate 1 MW electricity; about 11,000 labourers would be required, and the overall energy output/input ratios are high.

In the state of Gujarat, one of the poorest in India, a social forestry programme has been implemented with village plantation schemes and roadside and canal-bank plantings. Free extension services and seedlings and the involvement of local councils, farmers, schools, etc., have helped reverse the worsening fuelwood problem of only a few years ago.

Philippines (Ref. 32)

A feasibility study has shown that a 9,100 hectare fuelwood plantation 'would supply the needs of a 75 MW steam power station if it were not more than 50 km distant'. The investment requirements and cost of power produced looks favourable and competitive with oil-fired power stations of similar capacity. The best species of fast growing tree seems to be the 'giant ipil-ipil' (*Leucaena leucocephala*) which fixes nitrogen to ammonia - a highly desirable trait.

China (Ref. 38)

During the 1970's biogas plants have been perfected and installed at a rapid rate. It is estimated that there are now (1980) about 7 million in operation (about 1 million a year are being constructed) and the methane so produced is used for cooking, crop drying, power generation and various other purposes. In Szechwan province alone, 17 million people use biogas for cooking and lighting; in some areas 80% of all rural households are served by biogas. In addition over the last two decades a massive reforestation programme has increased the proportion of their national land under forest from 5 to 13% - implying an increase of 72m hectares.

Australia (Ref. 30)

In Australia, *Eucalyptus*, cassava (*Manihot*), *Hibiscus*, napier grass (*Pennisetum*) and sugar cane have been selected as potentially the most desirable high-yielding crops which can be harvested over the whole year. It was shown that alcohol produced from cassava (starch-rich) is an economically viable system, but that if processing to destroy cell walls is required the costs become too high. The cost of alcohol from cassava is calculated to be Australian $250/tonne from a 100,000 tonne/year batch-process plant which compares favourably with the current market price of alcohol (Australian $275/tonne) as an industrial solvent. Recent studies show that over half Australia's liquid fuel requirements could be produced from specifically-grown crops and agricultural residues.

New Zealand (Ref. 31)

With its very efficient agriculture, low population density and the fact that it uses a large proportion of its foreign exchange for petroleum purchases it is natural that New Zealand is considering biomass as a source of fuels. Current work shows that fodderbeet for ethanol and wood for methanol may be economic sources of liquid fuels. Proposals have been made that New Zealand could convert from a petrol to an alcohol liquid fuel economy - half provided by methanol from indigenous natural gas and the other half by alcohols from trees and agricultural crops.

South Korea (Ref. 12)

'The dramatic forestry success story of the seventies' is how the six-year-old South Korean reforestation scheme has been described. Since 1973, in the face of severe timber and fuelwood shortages, a community-based forestry programme has been implemented with priority being placed on meeting the needs of the rural population. Government statistics claim that one-third of the land is now stocked with trees less than 10 years old. A network of village forestry associations has been stablished providing a link between government organisations and the villagers - these associations establish, manage and harvest the plantations. Through profit-sharing mechanisms the Koreans have managed to co-opt private land for public forestry purposes. It has been calculated that a switch from coal purchases to locally grown fuelwood has meant a 15% increase in income for many families.

There is a resemblance between South Korea's and China's reforestation programmes. Central government in both cases had made a strong political commitment to community forestry with back-up technical assistance and local-level participatory institutions. Such patterns could be usefully examined in other so-called 'hopeless situations'.

In Table 6 the economics of a number of biomass energy products from various countries are summarised.

FUTURE PHOTOSYNTHESIS (Refs. 1,13 and 52)

One of the 'problems' with photosynthesis is that it requires a whole plant to function - and the problem with whole plant photosynthesis is that its efficiency is usually low (less than 1%) since many limiting factors of the environment and the plant itself interact to determine the final overall efficiency. Thus a task for photosynthesis of the far future is to try to select and/or manipulate plants which will give higher yields with acceptable energy output/input ratios. We need much more effort placed on studies of whole plant physiology and biochemistry and their interactions with external (environmental) factors. Already this type of research is being increasingly funded by both industrial and government organisations - in marked contrast to the neglect of this type of research in the past.

Examples of the type of research which is being done or needs to be done are: photosynthetic mechanisms of carbon fixation; factors in bioproductivity; genetic engineering using plant cell tissue cultures; plant selection and breeding to overcome stresses (drought, temperature or salinity); selection of plants and algae with useful products such as oil, glycerol, waxes, or pigments; nitrogen fixation and metabolism and its regulation by photosynthesis.

Also to be seriously considered is long-term basic, diverted research on artificial photobiological/chemical systems for production of fuels and chemicals (H_2, fixed C, and NH_3) needs sustained funding if these exciting future possibilities are to be realized.

CONCLUDING STATEMENT

Photosynthesis is the key process in the living world and will continue to be so for the continuation of life as we know it. The development of photobiological energy conversion systems has long term implications. We might well have an alternative way of providing ourselves with food, fuel, fibre and chemicals in the next century.

Suggested timetable of biomass-for-fuel programmes

Next 10 years: fuels from residues, trees and existing crops; use of existing biofuels; demonstrations and training.
10-20 years: increased residue and complete crop utilization, local energy crops and plantations in use.
After 20 years: energy farming; improved plant species; artificial photobiology and photochemistry.

ACKNOWLEDGEMENT

This article formed part of a study done by the author for a UNESCO study on 'Fundamental World Energy Problems', a UNEP project on 'Photosynthesis in Relation to Bioproductivity', and a UN Biomass Panel paper.

TABLE 1 Fossil fuel reserves and resources, biomass production and CO_2 balances (Ref. 1)

	Tonnes coal equivalent	
Proven reserves		
Coal	5×10^{11}	
Oil	2×10^{11}	
Gas	1×10^{11}	
	8×10^{11}t =	25×10^{21}J
Estimated resources		
Coal	85×10^{11}	
Oil	5×10^{11}	
Gas	3×10^{11}	
Unconventional gas and oil	20×10^{11}	
	113×10^{11}t =	300×10^{21}J
Fossil fuels used so far (1976)	2×10^{11}t carbon =	6×10^{21}J
World's annual energy use		3×10^{20}J (5×10^{9}t carbon from fossil fuels)
Annual photosynthesis (a) net primary production	8×10^{10}t carbon (2×10^{11}t organic matter)=	3×10^{21}J
(b) cultivated land only	0.4×10^{10}t carbon	
Stored in biomass (a) total (90% in trees)	8×10^{11}t carbon=	20×10^{21}J
(b) cultivated land only (standing mass)	0.06×10^{11}t carbon	
Atmospheric CO_2	7×10^{11}t carbon	
CO_2 in ocean surface layers	6×10^{11}t carbon	
Soil organic matter	$10\text{-}30 \times 10^{11}$t carbon	
Ocean organic matter	17×10^{11}t carbon	

These data, although imprecise, show that (a) the world's annual use of energy is only 1/10 the annual photosynthetic energy storage, (b) stored biomass on the earth's surface at present is equivalent to the proven fossil fuel reserves, (c) the total stored as fossil carbon on represents about 100 years of net photosynthesis, and (d) the amount of carbon stored in biomass is approximately the same as the atmospheric carbon (CO_2) and the carbon as CO_2 in the ocean surface layers.

TABLE 2 Solar energy for fuels: conversion process and products (Ref.1)

Resource	Process	Products	Users
Dry biomass	Combustion	Heat, electricity	Industry, domestic
(e.g. wood, residues)	Gasification	Gaseous fuels - methanol, hydrogen, ammonia	Industry, transport, chemicals
	Pyrolysis	Oil, char, gas	Industry, Transport
	Hydrolysis & distillation	Ethanol	Transport, chemicals
Wet biomass (e.g. sewage, aquatics)	Anaerobic digestion	Methane	Industry, domestic
Sugars (from juices & cellulose)	Fermentation & distillation	Ethanol	Transport, chemicals
Water	Photochemical/ photobiological catalysis	Hydrogen	Industry, chemical, transport

(Simplified table: numerous cross-links exist. Agriculture included in Industry. Many important final products not listed).

TABLE 3 Some advantages and problems foreseen in biomass-for-energy schemes

Advantages

1. Stores energy for use at will
2. Renewable
3. Versatile conversion and products; some products with high energy content
4. Dependent on technology already available with minimum capital input; available to all income levels
5. Can be developed with present manpower and material resources
6. Large biological and engineering development potential
7. Creates employment and develops skills
8. Reasonably priced in many instances
9. Ecologically inoffensive and safe
10. Does not increase atmospheric CO_2

Problems

1. Land use competition
2. Land areas required
3. Supply uncertainty in initial phases
4. Costs often uncertain
5. Fertilizer, soil and water requirements
6. Existing agricultural, forestry and social practices
7. Bulky resource; transport and storage can be a problem.
8. Subject to climatic variability

TABLE 4 Annual biomass production in tonnes

Net primary production (organic matter)	2×10^{11}
Forest production (dry matter)	9×10^{10}
Cereals (as harvested)	1.5×10^{9}
as starch	1×10^{9}
Root crops	5.7×10^{8}
as starch	2.2×10^{8}
Sugar crops	1×10^{9}
as sugar	9×10^{7}

TABLE 5 Biomass energy products costs compared to conventional - USA (1981 $) (Ref. 18)

Product	Cost from biomass ($/$10^6$ BTU)	Conventional cost ($/$10^6$ BTU)	Biomass/Conventional
Methanol	8.4 - 15.9	8.4	1.0 - 1.9
Ethanol	15.0 - 36.3	19.6	0.8 - 1.9
Medium BTU gas	4.7 - 7.4	3.0 - 5.0	0.9 - 2.5
Substitute natural gas	4.8 - 7.3	3.0 - 5.0	1.0 - 2.4
Ammonia	5.8 - 11.4	7.4	0.8 - 1.5
Fuel oil	3.6 - 7.9	3.2	1.1 - 2.5
Electricity	0.03- 0.14 ($/kWh)	0.03- 0.06 ($/kWh)	0.5 - 4.5

TABLE 6 Estimated biomass and energy product costs (various countries) (Ref.1)

	Country	Product and Source	Cost
1.	Brazil (1977)	Ethanol from sugar cane (ex distillery)	US$16.7/$10^6$ BTU US$ 0.33/litre
		Gasoline (retail)	US$13.8/$10^6$ BTU
2.	Australia (1975)	Ethanol from Cassava	AUS$250/t
		Industrial ethanol	AUS$275/t
3.	Canada (1975 & 1978)	Methanol from wood	CAN$0.35-0.70/gallon
4.	New Zealand (1976)	Ethanol from pine trees (500 t/day capacity; credits from byproducts)	NZ$260/t (13% return on capital)
5.	New Zealand (1977)	Biogas from plants	NZ$3.46-5.57/GJ
		Natural gas prodn. cost	NZ$1.09/GJ
		Coal gas prodn. cost	NZ$6.33/GJ
6.	Upper Volta (1976)	Fuelwood from plantation	US$0.09/kWh (t)
		Kerosene (retail)	US$0.13/kWh (t)
		Butane gas (retail)	US$0.11/kWh (t)
		Electricity	US$0.19/kWh (t)
7.	Philippines (1977)	Electricity from Leucaena fuelwood-fired generating station (same cost as oil-fired station)	US$0.014-0.018/kWH
8.	Tanzania (1976)	Biogas from dung (for cooking and lighting)	US$0.012/kWh
		Electricity	US$0.113/kWh
9.	India (Tamil Nadu) (1978)	*Casuarina* fuelwood to replace coal-fired electricity generating station (Competitive with coal: 15-30 year payback)	US$12/t(dry)

REFERENCES

1. Hall, D.O. (1978). Solar energy conversion through biology - could it be a practical energy source?. Fuel, 322-333; (1979). Solar energy use through biology - past, present and future. Solar Energy, 22, 307-328; (1979). Nature, 278, 114-117. Hall, D.O. and J. Coombs (1980). Biomass-growing as an energy source. In 'Must the world run out of energy?' Crown Agents Publi, London, S.W.1., U.K.
2. Eckholm, E. (1976). The other energy crisis: firewood. Worldwatch paper No.1., Worldwatch Institute, 1776 Massachusetts Ave., NW, Washington, DC 20036.
3. Mitsui, A., et al.,(eds.) (1977). Living Systems as Energy Convertors. Academic Press, N.Y.
4. Buvet, R., et al., eds. (1977). Living System as Energy Convertors. Elsevier, Amsterdam.
5. European Seminar on Biological Solar Energy Conversion Systems. (1977). UK-ISES, 19 Albemarle St., London. W1; also (1976). Solar Energy: a UK Assessment. Chap. 9.
6. Bio-energy: Energy from Living Systems (1978). Gottlieb-Duttweiler Institute, Ruschlikon, Zurich, Switzerland.
7. Schlegel, G., and J. Barnea, eds. (1976). Microbial Energy Conversion. Erich Goltz, Göttingen, Germany.
8. Earl, D.E. (1975). Forest Energy and Economic Development, Clarendon Press, Oxford.
9. Eighth World Forestry Congress (1978). Papers on 'Forestry for rural communities'. Jakarta, Indonesia, October 1978.
10. Arnold, J.E.M., and J. Jongma, (1978). Fuelwood and charcoal in developing countries. Unasylva, 29 (118), 2-9; Openshaw, K. and J. Morris (1978). Woodfuel - a time for reassessment. Natural Resources Forum, 3, 35-51.
11. King, K.F.S., and M.T. Chandler (1978). The wasted lands: the programme of work of ICRAF. Intl. Council for Res. in Agroforestry, PO Box 30677, Nairobi, Kenya.
12. Eckholm, E. (1979). Planting for the future: forestry for human needs. Worldwatch Paper No.26, Worldwatch Inst., 1776 Massachusetts Ave., NW, Washington, DC 20036; (1979). Forestry success of the seventies. New Scientist, March 8, 1979. pp.764-765.
13. St. Pierre, L.E., and G.R. Brown (eds.) (1980). Future sources of organic raw materials: CHEMRAWN I. Pergamon Press, NY. 651pp.
14. Department of Energy (USA) (1979). Mission analysis for the Federal fuels from biomass program. Vol.1., Summary and Conclusion. Publ. No. SAN-0115-T2. 64pp; Klass, D.L.,(ed) (1978). Symposium on Energy from Biomass and Wastes. Inst. of Gas Technology, 3424 South State St., Chicago, Ill. 60616, USA.
15. De Renzo, D.J. (1977). Energy from Bioconversion of Waste Materials. Noyes Data Corp., Park Ridge, New Jersey.
16. Biofuels: a Survey. (1978). Special Report ER-746 - SR, Electric Power Research Institute, Palo Alto, California.
17. Bio-Energy Directory Vol. 2. (1979). Bio-Energy Council, 1337 Connecticut Ave., Washington, DC; also Biomass Energy Institute, PO Box 1290, Box 129C, Winnipeg, Canada.
18. Flaim, S.J. and R.E. Witholder (1978). Policy aspects of biomass utilisation; Reed, T.B. (1978). Biomass energy - a two edged sword. SERI, Golden, Colorado. Papers presented at American Section of ISES, PO Box 1416, Killeen, Texas 76541.
19. Calvin, M. (1977). Hydrocarbons via photosynthesis. Intl. J. Energy Res., 1, 299-314; also (1977) Chemtech (June 1977), pp.352-362; also (1979) Petroleum plantations and synthetic chloroplasts. Energy, 4, 851-870.

20. Hammond, A.L. (1977). Photosynthetic solar energy; rediscovering biomass fuels. Science, 197, 745-6; Lipinsky, G.S. (1978). Fuels from biomass: integration with food and materials systems. Science, 199, 644-655; Burwell, C.C. (1978). Solar biomass energy: an overview of US potential. Science, 199, 1041-1048; Pimentel, D., et al., (1978). Biological solar energy conversion and US energy policy. Bio Science, 28, 376-382; Weisz, P.B. and J.F. Marshall. Highgrade fuels from biomass: potentials and constraints. Science, 206, 24-29.
21. Lewis, C.W. (1977). Fuels from biomass energy outlay versus energy returns: a critical appraisal. Energy, 2, 241-248.
22. Makijani, A. (1976). Energy Policy for the Third World. Intl. Inst. for Environment and Development, 10 Percy St., London WC1; Reddy, A.K.N. (1976). Non-commercial fuels in developing countries. Conf. C9. UK-ISES, 19 Albemarle St., London, W1. pp.79-82.
23. Vergara, W. and D. Pimentel (1978). A study of the energy potential of fuels from biomass in five countries. in Ref. 14.
24. W.M. Floor (1977). The energy sector of the Sahelian countries; also (1978). Energy options in rural areas of the third world. Policy Planning Section, Ministry of Foreign Affairs, The Hague, Netherlands.
25. Energy Needs, Uses and Resources in Developing Countries (1978). Report BNL 50784, National Centre for Analysis of Energy Systems, Brookhaven National Laboratory, Upton, NY 11973.
26. Inst. of Chem. Engineers (NSW, Australia) (1978). Alcohol fuels. Dept. of Chem. Engineering, Univ. of Sydney, ISNB No. 0-9596089-0-7.
27. Nathan, R.A. (1978). Fuels from sugar crops. US Dept. of Energy, Publ. TID-22781, Washington, DC 20545, USA.
28. Large scale methanol fuel production from surplus Canadian forest biomass: economic prefeasibility study. (1976). Environmental Management Service, Policy and Programme Development Directorate, Environment, Ottawa, Canada.
29. Canadian Renewable Energy News (1978). (a) Canadian Government Renewable Energy Programme 1978-1984, (b) Methanol from Biomass. Cren Publishing, Box 4869, Ottawa, Canada.
30. McCann, D.J. and H.D.W. Saddler (1976). Photobiological energy conversion in Australia, Search, 7, 17-23; (1979). Stewart, G.A., et al., CSIRO, Canberra, Australia. The potential for liquid fuels from agriculture and forestry in Australia. 146pp.
31. Stewart, D.J. (1977). Economics of biogas production from specifically grown plant material. Technical Report No.5, Invermay Research Centre, Mosgiel, New Zealand; also Whitworth, D.A. (1976). Production of liquid transport fuel from cellulose material (wood). NZ Energy J., 19, 173-177; also Harris, G.S. et al., The potential for energy farming in New Zealand, NZ Energy R & D Committee, Univ. of Auckland, Auckland, NZ. 2 vols. Report No.46.
32. Terrado, E.N. (wht P.V. Bawagan and J.A. Semana) (1978). in Sun: Mankind's future source of energy. ed. F. de Winter, Pergamon Press, NY.pp.101-105.
33. Groupe de Bellevue (1978). Projet Alter, Blvd. Port Royal 85, Paris 75013; also (1978). Science and Avenir (Paris). No.374, pp.18-23.
34. Sweden's sunny future (1978). Nature 273, 584; also Financial Times (London), European Energy Report No.8, May 23, 1978.
35. Goldemberg, J. (1978). Brazil: energy options and current outlook. Science, 200, 158-163; de Carvalho, A.V., et al., (1977). Energetics, economics and prospects of fuel alcohols in Brazil. CTP. Praia do Flamego 15, Rio de Janeiro 20000; Yang, U. and S.C. Trinidade (1979). Brazil's gasohol programme. Chem. Engin. Progress, April 1979, pp.11-19.
36. Brown, N.L. and J.W. Howe (1978). Solar energy for village development. Science, 199, 651-657; also (1978). Energy for the villages of Africa, Nat. Academy of Sciences, 2101 Constitution Ave., Washington, DC.

37. Seshadri, C.V. et al., (1978). Energy plantations - a case study for the Coromandel littoral. MC Research Centre, 11/12 North Beach Rd., Madras 600001.
38. Smil, V. (1978). China's energetics: a system analysis. Report for US Congress Jt. Economic Comm., Nov. 1978; also (1979). D. Lovejoy, United Nations, NY.
39. Bassham, J.A. (1977). Increasing crop production through more controlled photosynthesis. Science, 197, 630-638.
40. Wittwer, S.W. (1978). Future challenges and opportunities for agricultural and forestry research, see Ref. 13.
41. UK-ISES (1979). Biomass for energy. Conf. C-20, UK-ISES, 19 Albemarle St., London, W.1., U.K.
42. Watt Committee (1979). Energy from the biomass. Report No.5, Watt Committee, 1 Birdcage Walk, London, S.W.1. U.K.
43. Slesser, M. and C. Lewis (1979). Biological energy resources, Spon, London, 192pp.
44. Pimentel, D. and M. Pimentel (1979). Food, energy and society. Arnold, London, 165pp.
45. World Bank (1979). Prospects for traditional and non-conventional energy sources in developing countries. Working paper No.346, World Bank, 1818 H St., NW, Washington, D.C. 20433, USA.
46. Goldemberg, J. (1979). Global options for short-range alternative energy strategies. Energy, 4, 733-744.
47. Revelle, R. (1979). Energy sources for rural development. Energy, 4, 969-988.
48. Parikh, J.K. (1979). Renewable energy options: what could developing countries expect from them? Energy, 4, 989-994.
49. Commission of the European Communities (1979). Solar energy for development. Martinus Nijhoff, The Hague, 240pp.
50. Smil, V. and W.E. Knowland. (1980). Energy in the developing world: the real energy crisis. Oxford University Press, Oxford. 300pp.
51. Commission of the European Communities (1979). Energy R & D programme; second status report. Martinus Nijhoff, The Hague, 1055pp.
52. Bolton, J.R. and D.O. Hall (1979). Photochemical conversion and storage of solar energy. Ann. Rev. Energy, 4, 353-401.

SOLAR PROSPECTS FOR THE UK AND THE EEC

J. K. Page

Department of Building Science, The University of Sheffield, Sheffield S10 2TN, UK

The issue is not whether solar energy will work technologically, but rather when and where it will become of real value economically in Europe. Europe has a wide range of climates, and detailed analysis is needed of the typical patterns of solar energy supply and of energy demand in different areas before the true economic opportunities can be perceived.

Throughout Europe the energy used in buildings is about half the total energy consumption, so the role of solar energy in buildings is likely to remain a central issue. Climatic analysis demonstrates that, in general, north of the Alps solar heating prospects become better as one goes west. In the east the winters are very cold and relatively short. The Atlantic Seaboard in contrast has a mild winter climate followed by cool weather often extending well into the early part of the summer. The demands for summer cooling become more important as one moves south of the Alps, but the economics of solar cooling systems do not look favourable for the 80's. The systems currently proposed look too cumbrous to prove economic. It is an area in which a radical rethink is demanded. The very variable sunshine climate of Northern Europe demands special attention to the problem of thermal energy storage. The interest in passive solar buildings which store the solar energy within the fabric is likely to strongly increase, possibly at the expense of active solar systems. This will place the emphasis on the importance of methodologies of design.

Equipment will need to be made reliable to meet the market needs. Considerable progress has already been made in the development of EEC Standards for active solar collection systems. One can expect to see more complex systems developing, for example greater use of combined heat pump solar systems, combined wind/solar systems and so on. Prediction of probable performance will remain difficult. However, increasing use is likely to be made of computer simulation studies to predict the performance of such complex systems in advance of construction and to analyse the complex results of actual field trials. The performance of solar systems is more complex than many manufacturers have so far realised. A real effort will need to be made to reduce installation and control costs of active systems to reach cost effectiveness. It is likely that larger systems designed to provide a relatively smaller fraction of the total energy demand of bigger building complexes will prive more economic than small scale systems that attempt to supply a very large proportion of the energy from solar for an individual house, because the installation costs per unit area will be lower, the control costs do not increase relatively with scale and the system under-designed against the heating load will operate at lower collection temperatures giving inherently greater efficiency. The result will be a far better return on capital. Small may be beautiful, but medium size is likely to prove substantially better. Swimming pool installations are economically effective now and the immediate prospects in this sector are good. It is likely that a wider range of industrial applications, for example for temperature process heat, will emergy as economic. Once again scale will confer an advantage.

The compactness of Europe means there are few remote areas. The energy distribution infra structures are already well developed and the economic prospects for stand alone devices like solar cells will be less good than in less densely occupied and sunnier areas like Northern Africa. However, the recreational market for such devices is likely to become increasingly important, for example the boat battery charging market. The existence of relatively small markets are less important in luxury fields as the high overheads can be carried out of the profits which can in turn be used to help lower costs for other applications needing a greater inherent cost effectiveness.

Developments in biomass and central power tower electricity generation are dependent on land availability. Given the agricultural policy, an alternative to the subsidised overproduction of food could be the agricultural production of energy on a relatively massive sca scale. The chemical prospects for deriving a wide range of useful products from plants to replace petrochemical sources should not be under-rated. Our civilisation is hort of more than energy. Solar power towers are not likely to find a big role as such as lot of cloud occurs. The land take is large and is likely to be resisted in all areas where agriculture is possible. The EEC, however, could export hardware in this sector to more suitable areas.

As the real energy crisis in Europe is likely to rise in the 90's, the 80's will be a vital decade for long range scientific research in solar energy. The great strength of Europe in basic research is likely to prove a key asset and the scientific foundations for the world's energy supplies for the period 2000 - 2020 are likely to be established during the 80's with Europe making many important contributions.

The problem will be to convince governments facing increasingly grave short term financial problems of the vital significance of long range research. However, the long range future depends on the successful development of new industrial activity, therefore the real challenge is to industry. If one believes the future of resource supplies can be an extrapolation of the past, one can take on view of the future of industry. However, it appears increasingly likely that limited fossil energy resources available to the world will force another position. The energy gap is certainly too wide to be solved by nuclear policies. The industries that accept the need for change and take steps to assimilate the nature of this profound change into their manufacturing structure will be the industries that survive successfully into the 2000's. Solar energy could be the key to industrial success in Europe in the twentieth century and currently in Britain it is the long range industrial vision which is so lacking.

THE ROLE OF SOLAR ENERGY AND ENERGY CONSERVATION

R. Stewart, B. Murphy and J. Healey

Atmospheric Sciences Research Center, State University of New York at Albany, 1400 Washington Avenue, Albany, New York 12222, USA

ABSTRACT

The greatest contribution of solar energy to the handling of the worldwide energy situation has been one of encouraging conservation. As the concept of solar energy was being established, amateurs and professionals alike began calculating the energy needs per unit area of buildings, and what percentage of that energy need solar could provide. Only then did we realize that the energy use per unit area was so highly varied, so unbelievably wasteful, so impossible for economic solar systems to overcome, only then did an emerging solar industry turn to energy conservation first - and that was a change in direction that has permeated much of the energy industry, conservation first, then solar. This direction has, and will, aid the solar industry.

Several projects have been carried out by the Atmospheric Sciences Research Center (ASRC) which provide the basis for our present position and future planning on energy matters. These projects will be summarized in the final paper. They include:

1. The Alumni House - Conference Centre, State University of New York. The AHCC is a 740 m^2 building requiring 4200 kJ/^{o}C-hr for heating. A physical and economic analysis has been carried out. Over 60% of the building heat load was provided by the solar system in a 4200°C-day climate.

2. Photovoltaics at the 1980 Winter Olympics. The demand for photovoltaics is an international market which, it appears, will outstrip the capability of existing industry to satisfy. This should be a high growth area during the 1980's. The concept to be discussed involves three photovoltaic powered communication systems which have been installed in Whiteface Mountain, site of the 1980 Winter Olympics down-hill and slalom runs. These units, powered by an ARCO-Solar photovoltaic array, can support (1) a twenty-five station intercom for information distribution on the mountain, (2) a twenty-five station telephone system to control race communications, and (3) provide back-up for the communications console in the event of a major power shortage.

3. Energy conservation for local governments. A training and implementation program for local governments in the State of New York which decreased energy usage by 100,000 barrels of oil in the first year.

4. Energy Conservation/Solar Energy for Restaurants. The results of a three-year energy conservation program in a family style restaurant. A solar system, installed in the autumn of 1979 will be analyzed separately, and in comparison to the energy conservation system.

5. Minimum energy building. An analysis of a small polyurethane building designed without a conventional heating system. A stand alone solar panel was attached to aid in heating.

6. Heat pump technology - An analysis of present state of the art plus indicated directions of development for the 1980's. This analysis is based on the technical planning and outlook of several corporations.

KEYWORDS

Energy Conservation, Photovoltaics, Heat Pumps, Thermal Engines.

1. The Opportunity for Solar

On June 20, 1979, President Carter suggested a goal for solar energy systems by the year 2000. This goal was 18 Quads (1Q = 10^{15} BTU). Goals should be ambitious and the opportunities for the worldwide energy community which such goals offer should be explored.

Based on present day technologies and efficiencies the marketplace for the northeastern portion of the United States (PA, NJ, NY and the sic New England States) represents over six million installations of solar panels at an average size of 50 m^2 each. Based on a cost of 350 US dollars / m^2 this represents a marketplace of 100 billion US dollars.

The opportunity for wind machines is still being analyzed. Based on recent testimony in the US Congress (by Dr. James Lerner, September 24, 1979, a one-billion dollar, 500 MW program by 1987 must be considered as a minimal program. A commitment of 2-2.5 billion US dollars over the next seven years could result in the siting of 5000 MW by 1989. Recently the Government of Sweden awarded a contract in excess of 10 million US dollars for the design and construction of a 3 MW wind machine to Karlskronavarvet AB and Hamilton Standard (US). Such joint efforts by industry should develop the initial cooperative spirit so as to set the groundword for much larger contracts in the future. For the northeastern US, a range of wind machine sizes will be required. If we assume that 50% of the machines will be 5 kW, 25% will be 25 kW, 15% will be 200 kW and 10% will be 1 MW, then over 170,000 machines will be required at approximately 2,000 US dollars/ kW.

The photovoltaic market has been examined by several organizations. The year 2000 goal for the northeaster US will require over 37 million m^2 of cells, plus the needs of other sections of the US, and the rapidly growing international market. US firms have been forming liaisons with European firms, to pool technological resources and to establish international trade connections. Financially, the period of 1980-85 will see US firms involved in the international market as the cost of electricity vs the cost of photovoltaics does not yet support a large US market. The latter half of the decade should see a reversal in this process as the cost of electricity in the US rises to meet the decreasing cost of producing photovoltaic systems (electric rates in the US range from 2-11 cents/kWh at 1980 prices).

In addition hydroplants will be upgraded and new plants will be installed. Passive solar systems will require several hundred million m^2 of glazing plus innovative energy storage systems.

The year 2000 solar goal outlines a massive opportunity for the international industrial community. Whether or not the goal is reached, the process of reaching will create new employment opportunities for many skilled and semi-skilled workers, encourage research programs, and provide support for a host of related industries.

2. The Opportunity for Conservation

In the year 2010 "very similar conditions of habitat, transportation and other amenities could be provided" in the United States using twice the energy consumed today (1975 data) or alternatively using almost twenty per cent less than used today. (US Energy Demand, Science, April 14, 1979). Thus the present standard of living within the US could be maintained under a dangerously wide range of energy demands. Most importantly the lower end represents a 20 per cent decrease in energy usage. While energy trends take several years to develop a few examples will provide some concept on the impact of specific conservation measures.

Transportation consumes 25 per cent of the energy usage in the United States. Much of this usage is tied to the massive development of a highway system which reaches 5000 km in an east-west direction and 2250 km in a north-south direction. However, with automobiles averaging 21-23 km per US gallon in 1978-79, an increase to 45 km per gallon in 1986-87 will decrease total US energy usagy by 10-12%. No other single energy program can compete with the effects of this conservation program. Obviously, the weight of the 1978-79 automobile will have to decrease. The automotive emissions program will continue to be more restrictive, thus competing with the desire for better km/gal. However, the 1979 gasoline usage was 5 per cent below the 1978 usage, a first step. At present the international market controls many of the components installed in compact and sub-compact range of automobiles. The need for this size, and this weight automobile will increase. Developments which allow part for part substitution with lighter and less costly equipment is a growing market. Thus, there are several entry points for existing and new technologies in the US automobile market, especially if they aid in energy conservation.

Approximately 40 per cent of US energy consumption is contained in the conditioning of living space, i.e. lights, heating, air conditioning, hot water, etc. Case studies have shown substantial decreases in energy usage - and may point the way to the future choices of energy conservation methods:

(a) IBM reduced energy consumption by 39 per cent in four years ar 34 locations.

(b) The State University of New York at Albany reduced energy usage by 23 per cent in five years.

(c) The Marriott Corporation reduced energy usage by over 10 per cent in motels and restaurants.

(d) Many of the 67 county governments in the State of New York reduced energy usage in buildings by 10-20 per cent. This was based on a program initiated by the New York State Energy Office and resulted in a savings of over 100,000 barrels of oil in the first year.

The trend is obvious, the opportunities considerable. The process by which the above changes have taken place is similar. First an inventory of energy usage - an energy audit - takes place. This audit provides the necessary energy use profile, and economic implications, for decision making (for additional information see Landsberg & Stewart, 1980, Improving Energy Efficiency in Buildings: A Management Guide, State University of New York Press, Albany, NY, USA). Consulting firms provide most of the energy management usage profiles, or a more elaborate audit designed to outline present energy usage profiles, or a more elaborate audit designed to provide the economic data (payback and return on investment) which will allow for capital investment for longer term gains. Many firms now use computerized systems to carry out audits as well as to control energy usage as part of an energy management process. The computer industry has unlimited opportunities to provide both hardware and software as a normal part of energy management packaging. At present staff of the Atmospheric Sciences Research Center use microprocessors, minicomputers, data processors, and full size computers to carry out and control energy usage. Once energy auditing has been completed, retrofit of HVAC equipment may take place. One program, for schools, hospitals and local governments will expend over one billion dillars for auditing and retrofit. Other government programs, plus industrial retrofitting will more than double this amount within three years. Again, a substantial opportunity.

3. Energy Projects at the Atmospheric Sciences Research Center

3.1 Photovoltaics

The 1980 Winter Olympics at Lake Placid, NY, is the site for several photovoltaic demonstrations. Three systems are installed at the base ski lodge at Whiteface Mountain, site of the slalom and downhill events. Eighteen photovoltaic panels. provided free by ARCO Solar of Chatsworth, CA, charge twenty-four two-volt batteries (Figure 1). The batteries serve two communications systems: (1) the 48 v 2 amp race control point telephones which supports fifteen phones on Whiteface, a manual scoreboard locations, and a hand timing line in case of electronic failure, and, (2) a 24 volt 2 amp intercom system of 25-30 phones for continuous communication among Olympics officials. The third system is an emergency backup system which will support

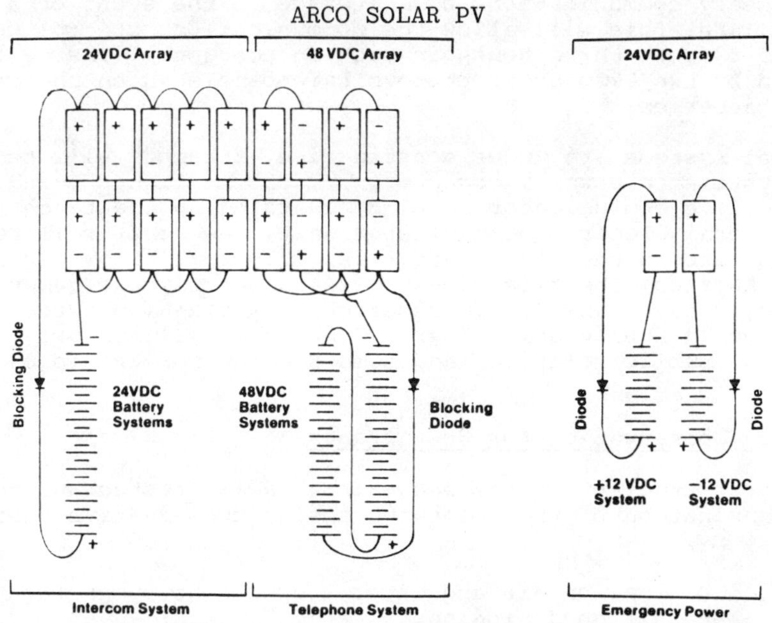

Fig. 1a Photovoltaic Solar Power System for 1980 Lake Placid Olympics

Fig. 1b SOLAR AT OLYMPICS: Solar electric photovoltaic cells are being installed at the Whiteface Mountain Ski Lodge near Lake Placid, NY by Arthur Rudin (left) of ARCO Solar, Inc. and Ronald Stewart of the State University of New York at Albany. The project, under the coordination of the USDOE Northeast Solar Energy Center of Boston, Mass., provided power to the communications systems for downhill ski events at the 1980 Winter Olympics.

all necessary communications on Whiteface in the event of a utility power failure. This will allow the communications control console to operate for up to three hours in such an emergency. The system is supported by two ARCO-Solar photovoltaic panels which charge two 12 volt batteries.

Additional systems are under construction. Attempts will be made to use PV systems to power a computer, TV, clock, lights - and a hand-held satellite communicator unit. By charging up the battery system of the General Electric communicator unit, the staff will be able to communicate to the ATS-3 satellite. This will allow discussions with Los Angeles, the site of the 1984 Summer Olympic games, and the Antarctic, site of ASRC field research in world-wide aerosol climatology. The Northeast Solar Energy Center (USDOE) has provided co-ordination, public relations and some funding for this project.

3.2 Energy Conservation for Restaurants

A three year study of energy usage in a family restaurant resulted in the examination of the following energy conservative features (Figures 2 & 3):

(a) Combination air and water source heat pumps for heating and air conditioning,

(b) Heat reclamation systems for the air conditioning units, the refrigeration rack, dishwasher, and cooking line exhaust,

(c) Heat pump - heaters that reclaim heat from the cooking line exhaust and heat water with an efficiency of as much as four times that of conventional electric water heaters,

(d) Thermal sotrage tanks,

(e) Controls for incoming and exhaust air based on real ventilation, and

(f) Controls for lighting level.

This restaurant decreased energy usage by 30% as compared to similar size restaurants. The best economic return comes from heat reclamation from the dishwasher, heat reclamation from the refrigeration rack and heat reclamation from the cooking line exhaust. (For additional information see Executive Summary: Jolly Tiger Restaurant/ Colonie, NY, Energy Conservation and Heat Reclamation Project, ASRC Report No. 678).

3.3 Minmum Energy Building

The Minimum Energy Building is a small 16 x 20 ft (4.8 x 6 m) two level structure. The insulation panels (polyurethane sandwiched between sheets of aluminum, or aluminum and sheetrock) which form the walls and ceiling are load bearing units. These panels form a somplete shell around the structure, under the basement floor, out-

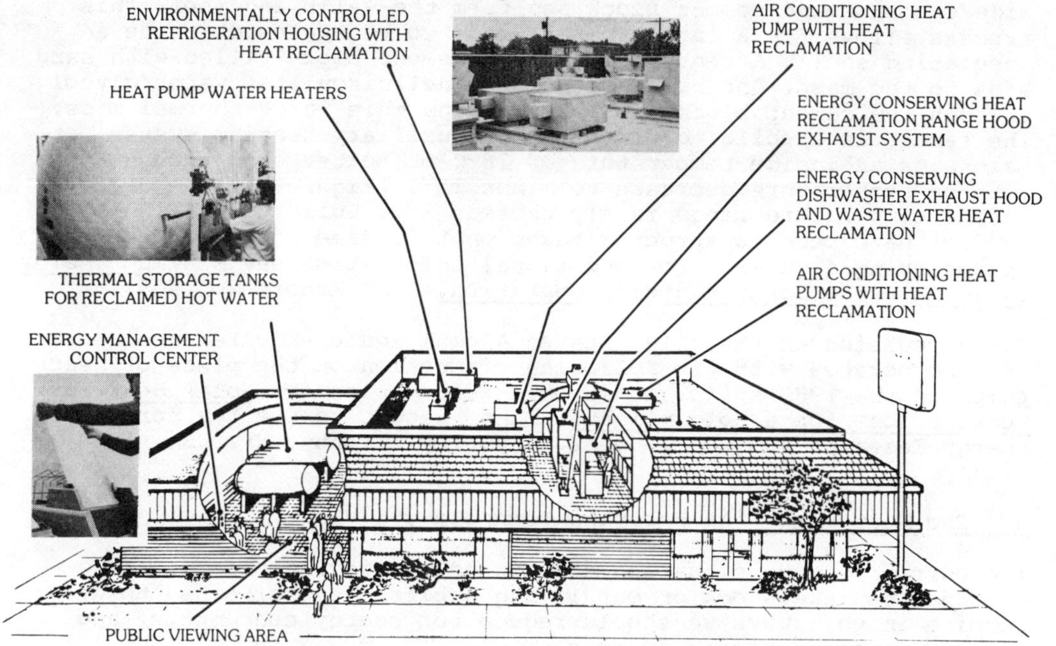

Fig. 2 Jolly Tiger Restaurant in Colonie, New York Incorporates a Hewlett-Packard 3050 Data Acquisition System to Collect and Analyze Data from Eight Energy Conservation Systems.

Fig.3 Photograph of a Cutaway Model of the Jolly Tiger Restaurant. This energy conservation/solar energy project is located in Colonie, NY, USA. The rooftop hybrid heat pumps are seen in the center of the photograph and the two storage tanks in the lower foregound.

side the basement cement block and form the walls and roof. This process allows for a large thermal mass to be enclosed inside an insulating shell. A central column of cement block filled with sand adds to the mass. One homemade solar panel circulates water/glycol through the cement blocks, slowly warming this large thermal mass. The result is a building which has no auziliary heating system yet maintains an inside temperature of 16°C or better even though the outside temperaturesdecrease to minus 25°C (Figure 4). The daily inside temperature range rarely exceeds 3°C. Buildings of 1000-5000 m^2 have been constructed based on this same process (see accompanying figures - for additional information see Minimum Energy Building - the First Winter's Operation, ASRC Report No. 405).

The discussion of the solar heated Alumni House - Conference Center is incorporated with the following discussion on the place of heat pumps in the 1980's (for additional information see Solar Heat Pump Combinations for a Cold Climate, Final Report to the New York State Energy Research and Development Authority, Albany, NY, USA, 1979).

4. The Heat Pump - An Evolving Technology

The current energy situation has caused us, as individuals and as nations, to take stock of our living habits. We are forced to determine in what ways we should reduce our energy consumption and harness untapped energy sources.

Depletable energy consumption may be reduced in either of two ways: (1) a cutting back in the total number of functions performed by non-renewable fuels or (2) searching for ways to reduce the amount of non-renewable energy required to perform a given task. Research is being increased in the search for added ways to harness untapped non-renewable fuels (such as coal gasification) as well as the safer, non-polluting renewable forms which include wind, hydro and solar power.

The heat pump is a device which incorporates both desirable solutions. It is capable of satisfying a given heating requirement using less non-renewable fuel. The source of heat which supplies the difference in work input is derived from renewable solar energy present in the surrounding atmosphere.

The American Society of Heating Refrigerating and Air Conditioning Engineers defines a heat pump, as it applies to year-round heating and cooling, as follows: "... a system in which refrigeration equipment is used in such a manner that heat is taken from a heat source and given up to the conditioned space when heating service is wanted, and is removed from the space and discharged to a heat sink when cooling and dehumidification are desired." [1]

The basic concept of the heat pump is to send a fluid (the refrigerant) through a closed loop. By means of expanding the fluid to its vapor state and then condensing it to its liquid state, heat is extracted from one high temperature space and deposited in a lower temperature space. The fluid flow is reversible, enabling a cooling effect to replace a heating effect. A diagram of a vapor compression heat pump depicts the basic components in the heat pump cycle.

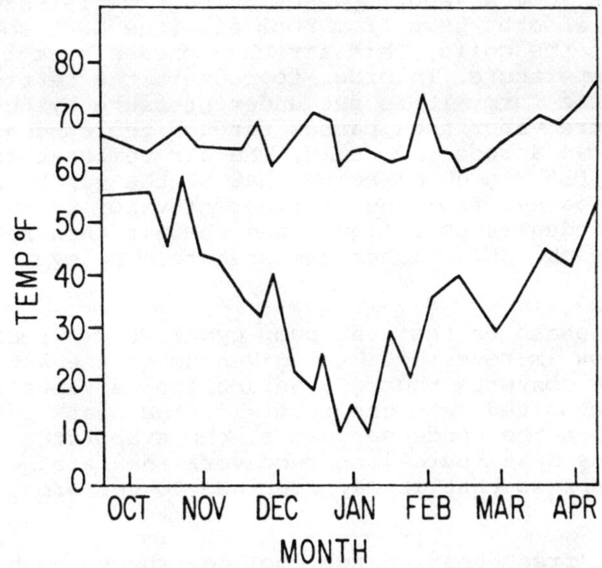

Fig. 4 Temperature History of Minimum Energy Building

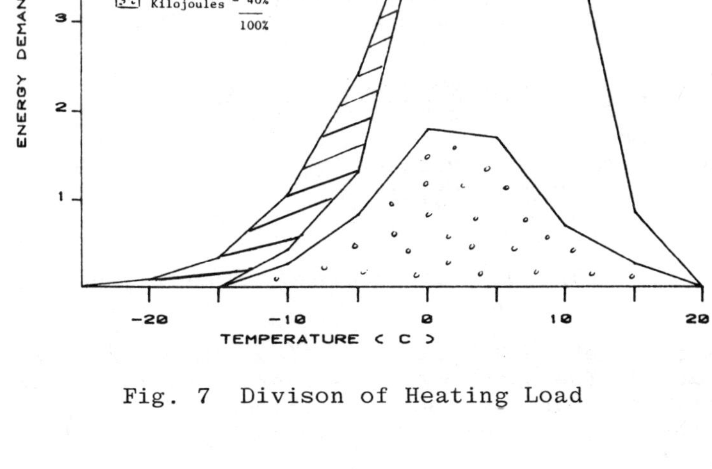

Fig. 7 Divison of Heating Load

In the cooling phase of the heat pump cycle, the refrigerant expands as it goes through the expansion valve and is cooled in the process. The refrigerant, now a vapor, passes through coils which are in the evaporator. It absorbs heat from room air (the heat source) which is blown across the coils. This air then passes back into the room at a cooler temperature. In order to convert the refrigerant once again to a usable form, it is put under pressure by the compressor. The high pressure vapor then passes through the condenser coils. Outside air is blown across the coils. The air temperature must be -9.4^{o} to 4.4^{o}C (15° to 40°F) below that of the vapor in the condenser. Excess heat is passed from the refrigerant vapor to the air. The cooled vapor condenses to a liquid and the air then returns to the outside (heat sink) at a higher temperature. The cycle can now be repeated.

In the heating phase of the heat pump cycel, the direction of the refrigerant flow is reversed. Heat given up by the action of the condenser as it converts the refrigerant from a vapor to a liquid. This heat is deposited into the room air (the heat sink) as it is circulated across the condenser coils. The evaporator uses heat from outside air (the heat source) to reconvert the refrigerant from a liquid to a vapor so that it may continue to perform its heating function.

The benefit of "free" heating (or cooling) which the heat pump is capable of delivering is best expressed by its performance factor - known commonly as its Coefficient of Performance (or COP). The COP is the ratio of the rate of heat delivered (or removed, whichever is the desired effect) to the rate of energy input to the heat pump system:

$$\text{COP} = \frac{\text{Refrigeration load}}{\text{Net work input}}$$

$$\text{COP} = \frac{\text{Net Heating (Cooling) Effect}}{\text{Net work input}}$$

$$\text{Heating COP} = \frac{\overset{1}{\text{Heat absorbed in evaporation}}+\overset{2}{\text{work of compression}}}{\underset{2}{\text{Work of compression}}}$$

$$= \frac{\text{Kilojoules}_1 \text{ \& Kilojoules}_2}{\text{Kilojoules}_2}$$

$$\text{Cooling COP} = \frac{\text{Heat absorbed in evaporation}}{\text{Work of compression}}$$

$$= \frac{\text{Kilojoules}_1}{\text{Kilojoules}_2}$$

Fig. 5 [3] Coefficient of performance (COP)

The formula for heating COP is an application of the First Law of Thermodynamics [4]: The quantity of heat produced is proportional to the work expended. For example, in heating a building, this law indicates that the thermal energy develivered to the building will be the sum of that extracted from a heat source, such as the outside air, plus the energy required to run the heat pump. Therefore, the total energy arriving in the area to be heated will always be greater than that required to run the device. Stated alternately, the on-site thermal efficiency, or coefficient of performance, will always be greater than one.

In terms of financial benefits, we pay for the kilojoules used to run the compressor, but the heat pump cycle freely provides the heating or cooling effect resulting from the change of the refrigerant between the liquid and vapor states.

In terms of the depletion of non-renewable energy, the heat pump's compressor consumes non-renewable energy (kJ_2) but delivers, on the average, twice as much as it consumes ($kJ_1 + kJ_2$). The source of the "free" energy in the heating phase is derived from ambient air heated by the sun's rays - a non-depletable resource.

The heat pump, a concept proposed by Lord Kelvin in 1852 [5], was initially introduced on the commercial market in the late 1940's (in the U.S.) [6]. Its major attraction at that time was in warmer climates where cooling was the primary air conditioning requirement. The heat pump's ability to reverse its cycle and become both a heating and cooling unit, was the feature which allowed its year-round use in regions with only occasional heating demands. In terms of cooling ability, the heat pump is comparable to a standard vapor compression air conditioner. In essence, it arrived on the market "ahead of its time". Its primary benefit of offering cheaper electric heating was not appreciated in an era of low-cost and abundant fossil fuels. In colder, norther climates, the reliability attainable with gas, oil or electric resistance heat could not be matched by the prototype heat pump models. Market forces were not sufficient to encourage the amount of research needed to introduce a higher level of reliability in heat pump performance.

In recent years, since the onset of the oil embargo, there has been and increased interest in the heat pump as an alternative or supplement to fossil fuels.

Capacity of the heat pump = its <u>ability</u> to <u>deliver</u> heat or cool to a building.

<u>Seasonal Performance Factor</u> (SPF) = value of the heat pump's <u>Coefficient of Performance</u> (COP) based on a <u>full season</u> of operation as well as the amount of auxiliary energy required.

Fig. 6

A look at the principles which govern heat pump operation reveals its potential strengths and weaknesses[7]. The "capacity" of the heat pump is its ability to deliver heat or cool to the building. The second law of thermodynamics expresses the concept that without an additional energy source, heat flows from a region of higher to one of lower temperature. In its heating mode, a heat pump appears to operate in the opposite direction. It extracts heat from a low temperature region, such as outdoor air at -1.1°C (68°F). The external energy added at the heat pump's compressor is the added source which pumps the heat "uphill". The larger the temperature difference over which the heat must be pumped, the greater the amount of external energy which must be supplied. The result is a decline in the heat pump's capacity, and in its Coefficient of Performance, as the season changes from moderate to extremes of hot or cold ambient temperatures. In other than moderate climates, the heat pump requires a supplemental heat source, usually electric heat with a COP of 1. Because of this temperature dependence, COP values based on a full season of heat pump operation are a more reliable guide to the system's actual performance. This seasonal COP, plus any required auxiliary energy, is referred to as the Seasonal Performance Factor (SPF). Due to such limitations as motor and compressor inefficiencies and requirements for fan power, SPF's to date have ranged from 1.5 to 2.5. Higher levels are theoretically possible. The cost at which improvements can be achieved are the primary determinants of their implementation.

Figure 7 indicates a breakdown of a typical heating load for a building utilizing a heat pump. The area under the top curve represents total building heating load required for a heating season (comparable to electric resistance heating with a COP of 1). The striped area represents the heat which must be supplied by the auxiliary heating system. The amount of kilojoules consumed by the heat pump is displayed under the bottom curve. The remaining clear area is the "free" heat captured from the outside air (or other heat source). In this instance, a Seasonal Performance Factor of about 2 is indicated.

The full potential for heat pumps as heating devices [8] has not been realized in northern climates because they were designed in the 1950's primarily for use in warmer regions. Heat pump optimization in a cold climate is obtained by choosing a balance point for the system components based on a <u>heating</u> "design season" rather than on cooling needs. (Balance Point is the outdoor temperature at which the capacity of the heat pump is just able to meet the heating demands of the building.)

Northern climates pose several specific challenges to heat pump performance. First, the evaporator is subjected to very low outdoor temperatures. This presents the need for frequent defrost of the outdoor coil unit. The typical defrosting method is to temporarily reverse the refrigerant flow, which converts the outside coil to a condenser. This procedure puts stress on the system components. It also requires more frequent use of the supplemental heating system. This lowers the seasonal performance factor and causes added expense for purchased fuel. A second major problem is the decline in reliability of the typical reciprocating compressor as temperatures dip below -10°C (30°F). This situation arises as a result of both a decreasing suction gas density and an increasing pressure ratio, against which the compressor must work.

A summary of the major reliability problems which plagued early heat pump models are identified in descending order of importance in Figure 8.

- Compressor failures
- Refrigerant leaks

Failures of:

- Outdoor and indoor fan motor
- Expansion valve
- Defrost system
- Reversing valve

Fig. 8 Major Reliability Problems

The picture just described may appear rather bleak for the consideration of the heat pump as a potential solution to the energy shortage. This is not the case. Heat pump state-of-the-art data are available from both government sponsored studies and heating and cooling industry publications. A review of this information reveals that during the '60's and '70's the technology has been developed which can correct the majority of the identified problems. The primary determinant of heat pump applicability is the cost of the improvements relative to the cost and availability of alternative heating devices and fuels.

Solutions which have been identified range from simple, obvious ones to others which are more complex and expensive:

- Place the compressor and controls indoors (for easier access, fewer cold weather problems)
- Provide a suction accumulator and oil return heat exchanger (to prevent liquid refrigerant from flooding back to the compressor)
- Base system design on heating season requirements (for higher efficiency in colder climates)
- Introduce multi-capacity compressors
- Improve defrosting system methods

Fig. 9 Recent solutions to early heat pump reliability shortcomings

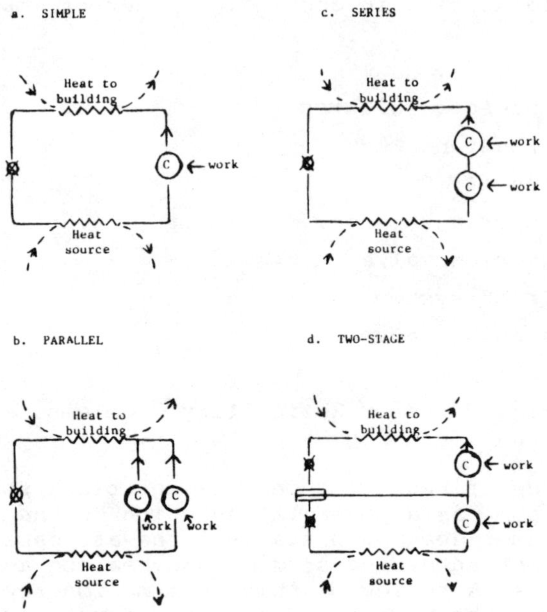

Fig. 10 Compression Cycles

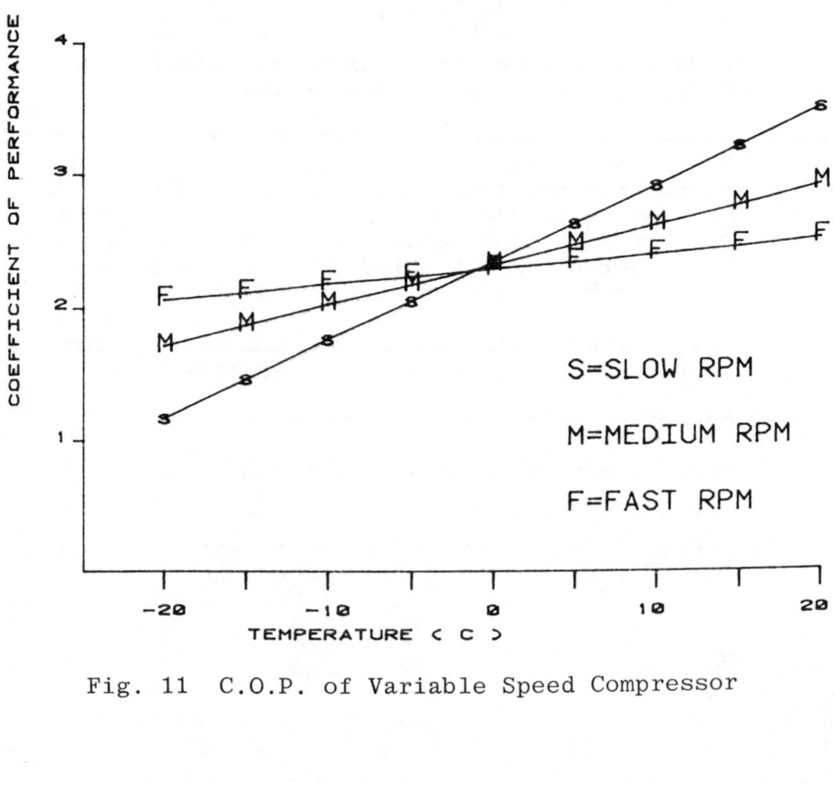

Fig. 11 C.O.P. of Variable Speed Compressor

Several modifications to the standard single-speed vapor compression cycle are possible at varying costs and levels of efficiency improvement[11]. (See Figure 10.) The simplest alternative is to use two parallel compressors. Each can be a different size, controlled independently to accommodate varying heating and cooling loads. This offers a better march between heat pump capacity (heating capability) and demand. A more even effect is achieved with two compressors placed in series. The lower pressure ratio across each compressor reduces adverse clearance volume effects which occur in reciprocating compressors. A two-stage compression cycle offers thermodynamic cycle improvements above those of the parallel and series approaches, further increasing the heat pump capacity and COP.

An additional increase in heat pump COP can be attained with the variable speed compressor, which can be controlled to a desired value of heating capacity. (See Figure 11.) A building's heating demand can be matched both below and above the original balance point temperature by increasing or decreasing the compressor revolutions per minute (RPM). Stress to the system is lowered because the ability of the heat pump to run more continuously eliminates the effects of frequent on-off cycling. Figure 9 indicates the effects of this compressor modulation on COP for a standard heat pump. When the heat pump is controlled to match the demand of a particular building at varying outdoor temperatures, it follows closely the locus of the most efficient operating RPM's.

Several improvements have been introduced to handle defrosting problems[12]. One method which came into use in 1976 (Janitrol) involves bypassing the condenser by injecting hot discharge gas directly into the evaporator. This eliminates cycle reversal but lowers efficiency. Another manufacturer (York) has introduced a solid state control panel, which among other functions, allows the heat pump to defrost on demand, based on temperature. Less efficient systems have a timed defroster which cycles on periodically whether or not frost is present on the outdoor coils. The temperature sensor results in a significant reduction in the number of defrost cycles, especially in dry climates.

Conventional heat pumps are air-source, reverse-cycle (Rankine) systems which employ electric resistance heat as a supplement at low ambient temperatures. Variations to the basic configuration have been introduced which are worthy of attention: Figure 10 lists the major variations.

- Hybrid Heat Pump
- "Heat Only" Heat Pump
- Absorption Cycle Heat Pump
- Thermal Engine Heat Pumps
- Water and Dual Source Heat Pump
- Solar-heat Pump Combinations

Fig. 12 Heat pump design variations

A Hybrid heat pump [13] is an electrically powered heat pump which uses a combustion furnace for supplemetal heat. This can be beneficial under certain circumstances. A combustion furnace utilizes scarce natural gas or oil. A hybrid add-on is worthwhile if the electric generation which powers it is accomplished using a fuel in more plentiful supply, eg. coal. Alternatively, widespread use of energy conserving systems such as solar or heat pump applications with electric back-up could accentuate the peak-load problem for electric utility companies during periods of extreme temperatures. A combustion furnace supplement has the benefit of reducing the winter electric utility peak-demand problem.

A heat-only heat pump [14] does not have a reverse cycle cooling mode. It's potential use is strong in much of Europe where air-conditioning is not a heavy requirement. Elimination of cooling design considerations permit the attainment of higher heating COP's (eg. COP of 2.3 instead of a conventional rating of 2.1 at -8.3°C (17°F) and 3.2 instead of 2.8 COP at 8.3°C (47°F). Heat-only heat pumps are available as add-ons to a resistance or combustion furnace.

An absorption cycle heat pump [15], in brief, is one which employs a secondary fluid, an absorbant, to absorb a primary fluid, the refrigerant. A generator replaces the compressor as the vehicle used to increase the pressure of the refrigerant prior to condensation. Evidence is as yet incomplete as to whether advantages will outweight the disadvantages of this heat pump mode. Each are enumerated in Figures 13 and 14.

- Inherently simple (few moving parts and potentially highly reliable equipment.
- High heating mode efficiencies.
- Lower concomitant energy requirements than thermal engine heat pumps.
- An absorption cycle generator can be more easily energized by a solar thermal source than can a thermal engine.
- On-site waste heat is available for evaporation defrost.
- Energy can be stored in the form of high pressure refrigerant vapor.

Fig. 13 Advantages of absorption cycle heat pump

- Relatively large, thus costly, heat exchanger area.
- Lower cooling mode efficiencies than those obtainable with the standard vapor compression system.
- Ammonia toxicity requires an intermediate means of transferring heat from the refrigerant to the conditioned space.
- Steel must replace aluminium or copper construction materials due to aqua-ammonia corrosiveness.
- Equipment is bulky.

Fig. 14 Disadvantages of absorption cycle heat pump

Absorption cooling is available in large and small size equipment. The absorption cycle unitary (factory assembled) heat pump is still in the design stage.

A thermal engine [16] is capable of replacing the electrically driven motor of the standard Rankine cycle compressor. Several configurations of thermal engine heat pumps are in the developmental stage. An on-site power source, such as a gas or oil-fired motor, enables engine waste-heat to supplement the refrigeration cycle output in the heating mode, and to assist in evaporator defrosting when necessary. This substantially increases on-site heating efficiency. Heat pump capacity modulation can be accomplished by varying the amount of fuel sent to the combustor. A thermal engine heat pump can be operated independently of utility lines; it is capable of being adapted to total energy systems.

There are negative aspects which apply in varying degrees depending on the particular type of thermal engine. For instance, waste heat becomes a detriment in the heat pump cooling mode. The higher level of system complexity may lead to greater cost and diminished reliability. On-site noise and pollution are increased. Finally, it is more difficult to design a totally hermetic (airtight) refrigeration system. While thermal engines may be driven by solar energy, most thermal engine applications will be dependent on scarce fuels such as natural gas and oil. This can increase dependence on these fules if it replaces energy conservation or alternative energy reliant on more plentiful sources. Alternately, this can cause lower total fossil fuel consumption to the extent that the thermal engine heat pump replaces a fossil fuel furnace. There are prototype thermal engines now being tested. For example, the Sterling-Rankine engine is being studied for heat pump application. It has among its benefits these features: a vertical self starting engine, no gears, noise free, and a potential for solar radiation to replace the combustor.

Heat pump design can be varied according to the "free" heat source which it utilizes [17]. The standard unitary heat pump in its heating mode draws upon a nearly inexhaustable supply-outdooe air. The heat pump's evaporator can extract heat from ambient air at temperatures as low as -6.7°C (20°F). The primary drawback of an air source heat pump is, as mentioned earlier, the need to switch to supplemental heat as the temperature decreases. A heat source which can minimize the use of supplemental heating is water. A water source heat pump is readily adaptable to use with a water storage tank and can be used in the operation of solar heating systems or electric peak shaving systems. When available and practical, lake, river or well water can be utilized as a heat source.

In solar heating and cooling system applications, a dual source heat pump [18] can make optimum use of both air and water. The water source is employed when the storage water temperature exceeds ambient air temperature in the heating phase. The air coil is brought into operation for milder heating needs and in the cooling mode, to reject heat to air rather than storage tank water.

Other less common sources are or can be derived from waste heat. A sewage-to-water heat pump began operation in Nuffield College in England in 1961. It ran about 11,000 hours in its first 10 years of operation. Heat pumps can be incorporated into total energy schemes such as cogeneration (the simultaneous production of electrical power and useful thermal energy) and district heating (locating dwellings and industry close to power plants to make use of rejected heat). Surface waters which maintain high temperatures due to thermal pollution released by industrial or generating plants can be tapped.

Researchers have been involved in extensive efforts to harness the energy of the sun. These efforts intensify as the fossil fuel supply crisis continues and as air pollution problems persist. The research efforts are proving successful; a solar system can be designed which is capable of supplying virtually all the heating and hot water needs of residential or commercial buildings. The principle barrier to such configurations is the system's size - hence, its prohibitably high initial cost. In a region where alternative fuels are scarce or their costs are high, a compromise results in a system which might satisfy approximately half of the building's heating load. Operators of electric utilities have expressed concern over the demand upon generating capacity which widespread solar system use will impose. They foresee the need to expand capacity to satisfy the electric load during periods of cold cloudy weather. (However, most U.S. utilities have a summer peak load now so that extra capacity for winter still exists.) This peak-load capacity will sit idle much of the time and its high construction cost will add an excessive burden to the rate payers. The heat pump has been tapped as a potential solution to this peak-load utility problem.

- Phoenix House Study: "Preferred Systems" = Solar + Storage + Heat Pump.

- Syracuse University Study: Solar/Heat Pump Combinations = wider range of usable temperatures than solar alone.

- Pennsylvania Power and Light Study: Dual source heat pump + Solar + Waste heat for storage = Total building heat load.

- Atmospheric Sciences Research Center Study: Parallel Solar-Heat Pump combinations = highest on-site energy savings.

- Solar Energized Heat Pumps - under study.

Fig. 15 Solar-Heat Pumps combinations

A study was completed in March of 1978 for a Colorado Springs municipally owned gas and electric utility company ("Phoenix House: Solar Assisted Heat Pump System Evaluation"[19]). The conclusion [20] was to recommend encouragement of "preferred systems". Such heating/cooling systems would be solar with seasonal storage in series with a heat pump. This combination proved superior of the various systems tested to minimize any detrimental impact on the utility load factor (ratio of actual mean electric generating load to maximum production capacity in a given period). This "preferred system" consumed only one-third the energy of its closest competitor - an electric resistance heating system with a load-management storage tank [21]. Other heating/cooling systems tested include electric, solar with electric backup and heat pump with electric backup.

Findings of a Syracuse University study in 1975 summarize general conclusions regarding the benefits of a solar assisted heap pump system: [22]

- Direct solar heating systems require water temperatures of 37.8°C (100°F) or greater. The heat pump assist enables heat delivery from temperatures as low as 10°C (50°F).
- Less expensive single-glazed collectors are sufficient.
- The capacity of the water tank is enhanced because of the wider span of temperatures at which the heat pump enables it to operate.
- A water to water heat pump can produce domestic hot water at reduced electrical consumption.
- Solar collectors can be used as a dry cooling tower to reject excess heat during the cooling season.

A demonstration project was conducted by the Pennsylvania Power and Light Company between 1972 and 1975 [23] in order to test methods of meeting growing electric demands. A tightly constructed home was built which utilized a heat pump as its primary heating source. Extra heat exchangers pick up the building's waste heat from exhausted air (refrigerator, washer, dryder, etc.) and re-integrate it into the system through the water source heat pump. Some 45% of the waste heat is recovered. A relatively small configuration of solar panels provides off-peak storage. A second outdoor air-to-water heat pump operates only when the water source heat pump requires energy from outside air. The study concludes that a dual source heat pump aided by solar, wind, or waste could adequately provide all the heating and cooling needs of a building plus savings in domestic water heating.

The Atmospheric Sciences Research Center has been involved in an extensive monitoring of heating systems at the 743 m^2 (8,000 ft^2) Alumni House Conference Center on the campus of the State University of New York in Albany since 1977 [24]. A comparison was made of the performance and costs of solar and heat pump systems, separately and in various combinations, against an electric heating system.

Factors including annual costs to own and operate each system, combined with availability and cost of alternate fuels become relevant for consumers. Local climate is an added factor in determining the optimum size of each system. An extensive simulation of annual costs as well as performance was made both of these systems and their logical substitutes. The substitutes considered were oil and gas furnaces and solar-heat pump systems which relied on either gas or oil as an auxiliary to the primary heating source.

Heating System	Energy Consumed (kWh/year)	% Savings $(1-\frac{\text{Energy Consumed}}{65,000})$
- Electric	65,000	0%
- Solar	40,000	38%
- 20 Ton* Air Source Heat Pump	28,000	57%
- 10 Ton* Air Source Heat Pump	33,000	50%
- Solar Assisted Water Source H.P.	29,000	56%
- Parallel Solar Heat Pump		
- Solar/20 Ton* ASHP	16,000	75%
- Solar/10 Ton* ASHP	22,000	67%

* 20 Ton = 253,000 kJ/hr; 10 Ton = 127,000 kJ/hr.

Fig. 16 State University of New York Alumni House Conference Center Heating Systems

Findings indicate that air-source heat pumps with oil or gas as a backup fuel offer equivalent or even lower annual costs than conventional gas or oil heating systems prevalent in the northeastern region of the United States. On-site heat energy svings for these heat pump systems are in the 50-60% range. Parallel solar-heat pump combinations, which offered higher on-sight efficiencies (67-75% less fuel consumed), could only compete with the more expensive annual costs of electric systems if their installed costs could be brought to about $160/m^2$ ($15/sf) of collector area. An added benefit of the heat pumps with combustion furnace backup is the fact that their reliance upon gas or oil relieves an electric utility of a potential decline in its load factor. For example, the Alumni House solar assisted heat pump economics would improve considerably if the system could be applied to constant year round use such as commercial water heating applications (car washes, laundromats, etc.). Such use in the New York region could produce over three times the fuel cost savings which resulted from heating the State University Alumni House building and providing for its minimal water heating requirements.

A final category, Solar Energized Heat Pumps, use solar energy as the heat pump's prime energy source [25]. This could be accomplished in one of several ways: (1) to drive the refrigeration cycle machine (by evaporating the working fluid of a prime mover) or (2) to evaporate the refrigerant in the generator of an absorption cycle machine. These concepts are at present in the experimental stage for heating. They may be even more promising in their potential for using solar energy to run cooling machinery.

It is a common practice to refer to the "on-site" performance of heating and cooling systems. This procedure is worthwhile for comparing relative efficiencies of similar systems. However, a more accurate measure of energy conservation is one which considers the consumption of scarce, non-renewable fossil fuels at the source of their consumption. Primary energy (also referred to as resource energy) requirements for heating or cooling equal the sum of the energy used at the building plus energy needed to produce and deliver those kWh's to the building. (See Fig. 17).

The principal energy loss for gas or oil furnaces is in combustion. This plus some losses in shipping result in 35% lost energy[26]. Electric heating is highly efficient at the building site. However, there is a staggering 70%[27] of each kWh of fossil fuel lost in electric generation and transmission. An electric heat pump with a Coefficient of Performance of 1.5 to 2.5 is capable of multiplying the 30% useful electric energy by its COP for an end use efficiency of between 45% and 75%.

Also to be considered is the question of what fuel is being consumed to produce the electricity. In the United States in 1975, about 35% of all electric generation was produced by the consumption of scarce fuels (petroleum or natural gas)[28]. Because of its abundance in the United States, the use of coal is being increased from its 1975 level of approximately 50%. Hydro and nuclear and some geothermal comprised the remaining 15%. Therefore, the decision regarding an optimum heating system becomes one dependent not only on cost and seasonal performance factor, but also upon locally available options. For example, somewhat higher primary fuel consumption could be preferable if the fuel consumed in one system (such as coal or solar for electric generation) were more readily available than another (ex. oil for a combustion furnace).

Oil or Gas:			
Distribution Losses	=	5%	35%
Combustion Losses	=	30%	
Useful Energy			65%
Electric: (derived from fossil fuel)			
Generation Losses	=	67%	70%
Transmission Losses	=	3%	
Useful Energy	=		30%
Electric Heat Pump:			
Electric Useful Energy	=		30%
Heat Pump Gain (COP of 2.5)	=		45%
Useful Energy	=		75%

Fig. 17 Examples of Primary Energy for Heating (% Useful Heat per Kilowatt-hour Consumed)

Our study at the Atmospheric Sciences Research Center compared the seasonal primary fuel use of the heating system at the Alumni House Conference Center.

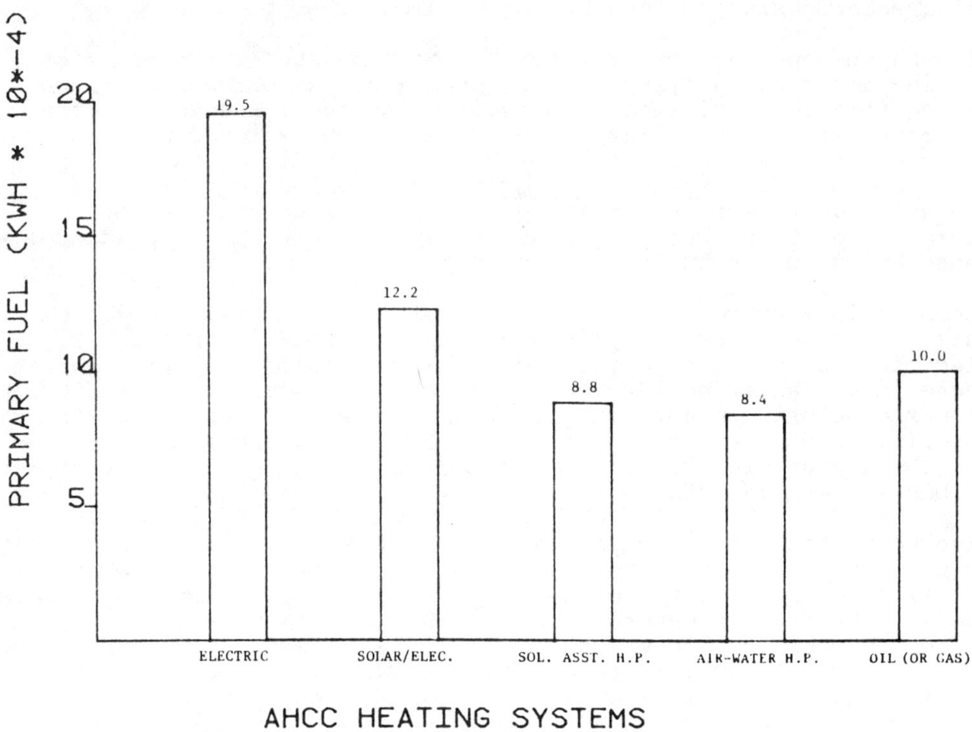

Fig. 18 Primary Fuel Use of AHCC Heating Systems

The air source heat pump (AWHP) with storage proved most efficient (16% savings over oil), followed closely by the solar assisted heat pump. An oil (or gas) furnace would consume less than a solar/ electric system. Electric consumed almost 60% more primary fuel than the heat pump systems. These heat pump configurations are commercially marketed products, their savings are available today.

A simulated comparison of heating and cooling configurations was performed as part of the Gordian Associates' Report to the U.S. Department of Energy in 1978. (See Fig. 19). Primary energy Seasonal Performance Factors of conventional equipment [29] were compared along with advanced technology which was then in the demonstration stage. Nine U.S. cities were included in these simulations. Two have been selected for our consideration because of their similarity with European climatic conditions. Seattle, Washington (47°N Lat) on the Pacific coast has conditions similar to the marine climate prevalent in much of Europe (accent on winter, with rain in all seasons). The climate in Concord, New Hampshire (43°N Lat) in the northeastern U.S. could be equated to those regions of Europe with humid continental climates (long winter snow cover prevalent). Heating rather then cooling is the principle concern in most of Europe as is the

	Heating/Cooling System	Seattle, WA			Concord, NH		
		Heating	Cooling	Heat & Cool	Heating	Cooling	Heat & Cool
1.	Electric Heat Pump-Standard	.59	.67	.60	.41	.57	.42
2.	Electric Heat Pump-High Efficiency I	.65	.71	.67	.43	.69	.45
3.	Electric Heat Pump-High Efficiency II	.68	.66	.69	.48	.61	.49
4.	Electric Heat Pump-High Efficiency III	.67	.69	.68	.46	.71	.48
5.	Hybrid Elec. H.P.-Gas Furnace	.52	.56	.53	.50	.53	.50
6.	Hybrid Elec. H.P.-Oil Furnace	.53	.56	.54	.52	.53	.52
7.	Electric Heat Pump-Heat Only/ Central Air cond.-High Efficiency	.75	.75	.75	.51	.64	.52
8.	Gas Warm Air Furnace/Central Air Cond.-High Efficiency	.55	.75	.56	.55	.64	.55
9.	Oil Warm Air Furnace/Central Air Cond.-High Efficiency	.61	.75	.62	.62	.64	.62
10.	Central Electric Furnace/Central Air Cond.-Heat Efficiency	.30	.75	.35	.28	.64	.30
11.	Baseboard Convectors/Window Air Conditioners	.32	.68	.34	.29	.67	.30

Fig. 19 Primary Energy Seasonal Performance Factors * for Currently Available Heating/Cooling Combinations - Residential

* High SPF is preferred.

Source of Data: "Heat Pump Technology, A Survey of Technical Development Market Prospects", Gordian Associates, 1978.

Heating & Cooling System	Seattle, WA			Concord, NH		
	Heating	Cooling	Heat & Cool	Heating	Cooling	Heat & Cool
1. Advanced Electric Heat Pump I	**	**	**	**	**	**
2. Advanced Electric Heat Pump II	.96	1.15	.98	.58	.97	.61
3. Free Piston Sterling Rankine Gas Heat Pump	1.09	.71	1.02	.98	.68	.93
4. V Type Single Cylinder Sterling Rankine Gas H.P.	1.11	.88	1.06	.91	.82	.90
5. Organic Fluid Absorption Gas Heat Pump	.94	.47	.83	.87	.46	.80
6. Pulse Combustion Furnace/ Central Air Cond.-High Efficiency	.82	.75	.81	.82	.64	.80

Fig. **20** Primary Energy Seasonal Performance Factors * for Advanced Heating/Cooling Combinations - Residential

* High SPF is preferred.

** System is not sized for application in these cities.

Source of Data: "Heat Pump Technology: A Survey of Technical Development Market Prospects", Gordian Associates report to U.S. Department of Energy, 1978.

case in the northern regions of the United States.

For residential applications, seven heat pump systems (#'s 1-7) which are available on the market were compared with standard electric resistence (#10-11) and fossil fuel furnace heating systems (#8-9). Of the seven heat pump configurations, one is a standard electric (#1) and three are high efficiency units (#s 2-4). A hybrid heat pump was tested with both gas and oil furnaces as supplements (#s 5-6). Finally a heat only heat pump was considered (#9). Electric resistence consumes the most primary fuel (exhibits the lowest Seasonal Performance Factor) of all. In Seattle's marine climate (similar to the eastern coastal countries of Europe) the high efficiency heat pumps outperformed oil and gas as well as electric resistence heat. The hybrid heat pumps had slightly lower S.P.F.s than oil or gas. The "heat only" heat pump outperformed all others in this climate (23% higher S.P.F. than oil furnace). In the colder continental climate of Concord (similar to mountainous inland European regions), electric resistence again exhibits the worst S.P.F. The "heat only" heat pump and the hybrid electric heat pumps came closest to the performance of the gas and oil furnaces.

Several advanced heat pumps were tested which are in the development stage for residential applications. (See Figure 20). Two (#s 1-2) are electric heat pumps which incorporate improved design features. Three others (#s 3-5) are heat pumps which operate (#6) is simply an advanced gas combustion furnace (no heat pump). Of these systems, all met or substantially outperformed the most efficient heating/cooling combinations of those tested which are marketed today.

Finally five commercial heating/cooling combinations were tested. (See Figure 21). Results were consistent in both locations. Of the currently available systems, the standard electric heat pump exhibited a higher S.P.F. than the two commonly used commercial installations - electric resistence and rooftop gas furnace with electric air conditioning. Both advanced commercial gas heat pumps indicate an even higher S.P.F. potential, with the steam Rankine-Rankine Gas Heat Pump exhibiting superior performance to the Bryton-Rankine.

Even as their report was being concluded in June of 1978, the authors alluded to the rapid developments occurring in the heat pump industry. These included manufacturers' planned marketing of a two-speed electric pump for residential applications and of dual-source, solar assisted heat pumps with thermal storage [30]. A perusal of trade association newspapers and journals over the last two years reveals the dynamic evolution of heat pump technology. Trade literature in 1978 and 1979 include among many ongoing developments, reports of new or improved models already installed or nearing market availability.

Heating & Cooling System	Seattle, WA			Concord, NH		
	Heating	Cooling	Heat & Cool	Heating	Cooling	Heat & Cool
1. Electric Heat Pump-Standard	.75	.70	.73	.59	.63	.60
2. Electric Resistence Heat/Electric Air Conditioning	.31	.71	.39	.28	.61	.34
3. Rooftop Gas Furnace/ Elec. Air Cond.	.43	.66	.49	.46	.59	.50
Advanced Commercial Applications						
1. Brayton-Rankine Gas H.P.	.93	.73	.85	.76	.72	.75
2. Rankine-Rankine Gas H.P.	1.35	.91	1.16	1.27	.91	1.11

Fig. **21** Primary Energy Seasonal Performance Factors *
Currently Available Heating/Cooling Combinations - Commercial

* High SPF is preferred.

Source of Data: "Heat Pump Technology, A Survey of Technical Development Market Prospects", Gordian Associates, report to U.S. Department of Energy, 1978.

- Coleman	- add-on gas furnace
- Carrier	- 2-speed heat pump - room size heat pump
- York	- demand defrost systems - high efficiency compressor - variable balance point setting - solar powered heat pump motor for heating and cooling - inverter drive for electric motor speed control
- Cam-Sat	- "De-Ice Control"
- Westinghouse	- heat pump waste heat reclamation - solar cost reduction with heat pump use
- Advanced Energy Technology	- "solarbank" heat pump for cold climate applications

Fig. 22 Manufacturers Announce Heat Pump Technology Advances: 1978-1979

The Coleman Company announced a heat pump add-on to a gas furnace [31]. Controls choose between operating with the gas furnace at temperatures below -3.9°C (25°F), with the heat pump at temperature above 8.3°C (47°F), or a combination of both between -3.9°C and 8.3°C. Heating COP is rated at 3.0 at 8.3°C (47°F), 2.3 at -8.3°C (17°F) and 1.5 at -23.3°C (-10°F). In its cooling mode, an Energy Efficiency Ratio of between 8 and 9 is quoted [32]. (EER is a rating for a heat pump's cooling capacity. It is a ratio of BTU's/hour divided by electricity consumed, measured in watts (1 BTU = 1.055 kJ = .293 wh). Seven is considered a good EER rating. Eight or higher is excellent[33])

The Carrier Corporation also manufactures an add-on heat pump system which is capable of operating in conjunction with electric, oil or gas central air conditioning with some modifications [34]. It claims as well a heat pump model with a 2-speed compressor motor [35]. When demand drops at moderate temperatures, a more energy efficient low speed is automatically switched on. A 1979 addition to Carrier's line is a small, room-size heat pump - a window unit which heats and cools [36]. Its features include a solid state variable speed fan and an "Econo-Zone" which suggests a range of thermostat settings to provide greater efficiency of operation.

The York Division of Borg Warner Corporation claims it has initiated a demand-defrost system into its solid state logic module. This eliminates as much as 80% of the defrost cycles initiated in the time-temperature defrost method found in many heat pumps [37]. The demand-defrost method is based on the pressure differential across the outdoor coil and can distinguish between coil blockage and wind gusts. The decrease in defrosting time reduces the amount of building heating load which must be met by the auxiliary system. York indicates that this improvement coupled with a new high-efficiency compressor can add 7% to heating and cooling savings over previous York add-on models [38]. Selection of a temperature cut-off point for the

the compressor can be based on local fuel costs. This proper matching of the heat pump and its auxiliary heating system is claimed to save up to 30% in heating bills compared with other heat pump models which automatically shut off at the balance point temperature.

York has installed centrifugal liquid chillers (water-to-water heat pumps) in a 9,300 m^2 (100,000 ft^2) Minneapolis building [39]. The chiller motors are driven either by Rankine cycle solar energy or conventional electric. A storage tank will retain any fluid heated by solar collectors which is not needed immediately to drive the chillers. The chillers are estimated to supply 60% of the building's energy needs.

Borg Warner has announced a breakthrough in efforts to apply alternating current electricity to compressor motors [40]. The breakthrough in marketing this inverter drive after 20 years of study came with the development of low cost microprocessor controls. The inverter concept provides almost infinite electric motor speed variation. The controls eliminate energy waste by matching the speed of the heap pump motor precisely to the heating load. York reports that when this system is applied to its split-system heat pump, it improves the already high COP (3.1) by an additional 17%. (Exxon claims similar efficiency increases from a version called the Alternating Current Synthesizer (A.C.S.) system.) [41]

Still another solution to the heat pump's defrosting problem [42] has been offered by Cam-Sat, Inc., a Los Angeles firm. The devise, called "De-Ice Control" uses neither temperature, time nor pressure as its initiator. Instead, it relies on the electrical load of the heat pump's condenser fan motor. The device has been used on an experimental basis for problem heat pump units with reportedly favorable results.

Westinghouse Electric Corporation manufactures a high temperature heat pump which can utilize industrial waste heat to supply hot water and winter heating needs. In announcing an application of this heat pump in a tractor manufacturing plant, Westinghouse enumerated the benefits: elimination of waste treatment problems and associated fan horse power costs and cooling tower water requirements; waste heat is conserved and used in another process [43]. Westinghouse has applied this same high-temperature heat pump in combination with a solar energy collection system [44]. It claims that this union reduces by one-half the amount of solar collectors required to perform an equivalent heating function, thus significantly reducing the solar system costs. The system has commercial and industrial applications.

Advance Energy Technology Corporation has introduced a patent-pending "second generation" heat pump, the "solarbank" heat pump [45]. It feels that the "solarbank" corrects the heat pump's major cold climate shortcomings. The compressor is located indoors, therefore, all heat produced can be utilized as warmth. The heat pump makes use of <u>internal</u> building heat as its heat source. A small array of solar collectos sends heat to a sand bank for storage. The heat pump can produce and store heat, or alternately freeze ice for cooling during the local electric utility company's off-peak hours. Advance Energy Technology states that this system also eliminates the problem-plagued 4-way seasonal changeover valve, the need to change compressor speeds and defrosting requirements with their attendant auxiliary electric costs.

These few industry reports are indications of the many applications of improved technology occurring in the heat pump industry today. In the United States, much of the heat pump research has been carried on by the HVAC (heating, ventilating, air conditioning) industry. Studies which are government and utility company funded have been added to these efforts in the interest of energy conservation. Realizing its energy conserving potential, both agencies have recently included the heat pump in various incentive packages.

- Proposed Tax Credit
- DOE contracts for heat pump-centred community energy systems
- Tennessee Valley Authority loans
- Long Island Lighting Company tests solar-heat pump homes

Fig. 23 Government and Utility Company Incentives for Heat Pump Installations

At present, legislation is pending in the U.S. Senate which includes a tax credit of up to 1000 US Dollars for installing residential heat pumps in place of less energy efficient systems [46].

The U.S. Department of Energy recently awarded seven contracts for the development of integrated community energy systems which feature heating and cooling in an area through use of heat pumps [47]. Non-scarce fuels, which would not be feasible in small systems, would be employed to power the heat pumps.

Electric companies which are part of the Tennessee Valley Authority are offering low- or no-interest loans to customers who use electricity for heating or cooling [48]. Among other conservation goals, the loans are aimed at encouraging the conversion of existing electric heating systems to more energy efficient heat pumps.

The Long Island Lighting Company, a New York State Utility company, is hosting a project in which five homes in its service area will be living laboratories for solar electric heat pump systems [49]. This is one of several company projects designed to test new ways of using existing energy resources more efficiently and economically. Another phase of the company's demonstration projects involves energy storage systems for heating and cooling.

Extensive heat pump research advances and the start of heat pump incentive plans are evident. Where will these efforts lead? Government and HVAC industry's studies have attempted to answer this question through analyses of the heat pump's market potential. Market projections are dependent upon various factors which affect the public's acceptance of this heating system.

- Cost and reliability of heat pump
- Cost and reliability of alternative heating systems and energy sources
- Current financial picture
- Consumer attitudes

Fig. 24 Factors Affecting Future Heat Pump Applications

The purchase price of a standard heat pump is in the range of costs of conventional heating and cooling systems in the U.S. [50, 51]. High efficiency models (now available) and advanced models (prototypes) may be expected to be 5 to 40% more expensive [52]. A more accurate guage of total heat pump system cost includes annual operating costs over an estimated 9 year life. These include the costs of maintenance and of auxiliary fuel. Maintenance costs become lower as general heat pump improvements come on line and as operating parts are protected from the weather or placed inside the building. A 1977 survey of heat pump contractors was conducted by a trade publication [53]. It revealed a low incidence of compressor problems in severe climates such as North Dakota and Minneapolis with the use of reliable brands (e.g. Carrier, G.E., York). These cities experience long, snowy winters comparable to inland European regions which have a "continental climate " [54, 55]. Proper mounting (above the "snow line") and installation (protection from wind, dripping water, etc.) have corrected or forestalled defrost problems. Heat pump installation is recommended only in conjunction with adequate insulation which is the first line of defense against consumption of auxiliary fuel. (A newly introduced Swedish ventilation system, the "Kantherm" is a potential solution to the problems related to the "superinsulated house": stale air, odors, excessive moisture and biological growth [56].) Unexpectedly high use of electric auxiliary heat was reported as a problem in some cases in the survey. The hybrid heat pumps now on the market would permit selection of the least expensive supplementary fuel available in a particular location. A heat-only heat pump would permit optimum capacity adjustments for regions which experience prolonged winters.

A dominant factor affecting heat pump acceptance is the cost and reliability of alternative heating/cooling modes. Central heating systems using gas, oil or electric resistence have been noted for high reliability. Installed costs are reasonable and system maintenance not excessive. Detriments to these systems are the low primary energy efficiencies for electric heat and the actual or potential scarcity of oil and gas. Heat pump efficiencies become more attractive where electric rates are high and as the price of oil rises at an alarming pace. Where gas is available and its price artificially suppressed, the heat pump may not yet be able to compete with a gas furnace. The potential for development of a high efficiency pulse gas furnace would offer further competition to the heat pump. However, fossil fuel availability is becoming more unreliable. Electricity for heat pump assist can be provided by locally available energy sources such as coal or hydro. The price of gas has been forecast to rise more rapidly than that of other fuels in the United

States [57]. These conditions are causing the reliability factor to tip in favor of the heat pump. Innovative applications of local resources can reduce reliance on supplementary fuels. A Canadian homeowner uses a nearby river as the heat source for the home's heat pump system. The relatively steady water temperature, combined with heavy home insulation and triple-glazed windows, reduced the estimated heating bill by two-thirds - from $1200 to $400 per year [58]. Other heat sources such as a building's internal waste heat in conjunction with a heat storage bed can also reduce the requirement for supplementary fuel.

The general financial climate has its effect upon potential heat pump installations. The availability of mortgage money and its current rate of interest affect the number of building starts in a given year [59]. New housing and commercial developments offer a high potential for additional heat pump use. High inflation is a two-edged sword in relation to heat pump acceptance. Government efforts to curb inflation have led to tighteining of mortgage money with higher interest rates - hence fewer housing starts. However, inflation is being "fueled" primarily by higher energy costs. This is causing consumers to consider all potential alternatives for energy savings with greater care. Life cycle costs of heating systems are replacing installation costs as a primary consideration. The heat pump is gaining a more prominant position because of its high fuel efficiency.

A final, but crucial factor affecting heat pump acceptance is consumer attitudes. Studies [60] of real estate sales indicate high acceptance of a heat pump in regions where the system is fairly well known, such as in southern United States. More educational efforts are needed where the heat pump is generally not known. Promotional efforts by HVAC contractors and utilities are occurring with increasing frequency and with noticeable results [61]. Potential residential building owners have indicated a willingness to invest in an energy saving system with a higher installation cost of the payback period is six years or less. Commercial investors seek a payback in about half this time. Currently, the six-year period is reasonable in many instances while the shorter payback might not be. This situation is changing constantly as the rate of increase in fuel costs accelerates and as fuel supply reliability becomes more strained. European countries have exhibited a more frugal use of energy to date. The pinch of high energy costs has been felt over a longer period of time. Reintroduction in the United States of concepts such as district heating [62], compactly built homes and heavy insulation are well established in many parts of Europe [63], especially where the housing and industrial building stock have been replaced since the 40's. Last year's record cold weather caused Europeans to stretch their imaginations for heating solutions[64]. Such energy consciousness lends itself to a growing acceptance of the efficient heating available through heat pump use.

An assessment of the current heat pump market sheds some light on the extent of its future penetration in the heating/cooling market. (See Figure 25.)

Figure 25 reveals that from 1967 to 1975 annual heat pump sales per year in the United States doubled from approximately 75,000 to 150,000 units [65]. Gas is the preferred heating fuel in the country. Oil furnaces were installed more extensively than electric until 1972.

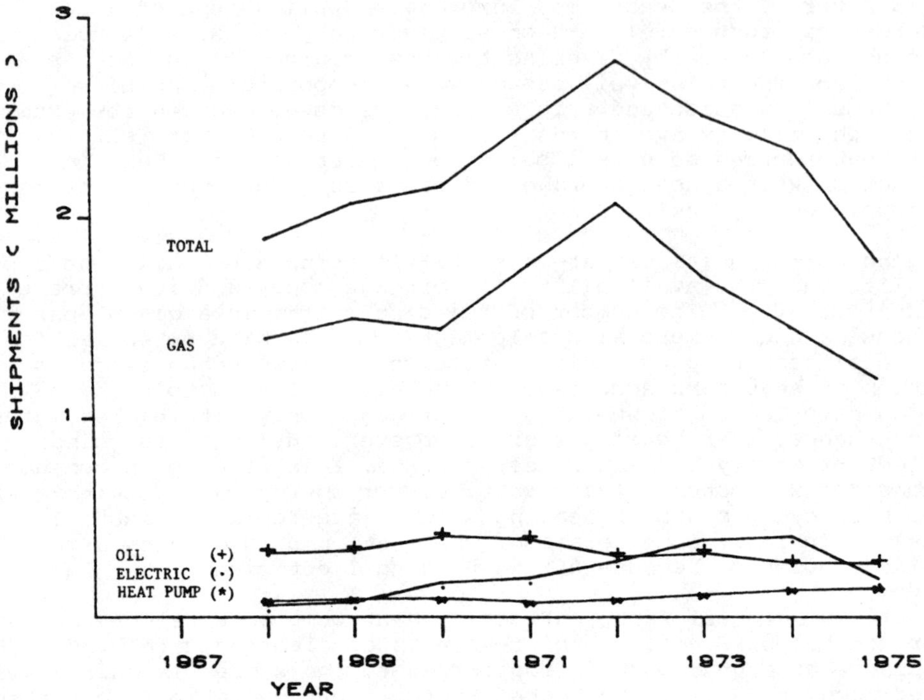

Fig. 25 U.S. Central Heating Equipment Shipments

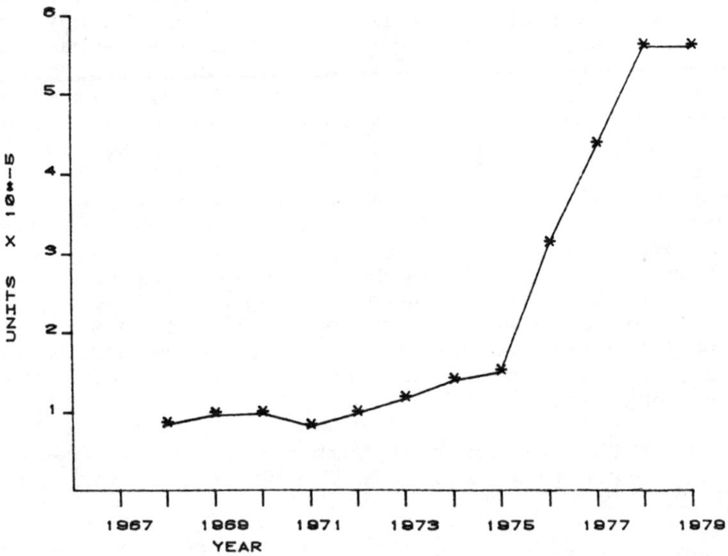

Fig. 26 U.S. Heat Pump Shipments

The heat pump claimed under 5% of all installations at that time.

Figure 26 extends the United States heat pump sales figures to 1978. Annual heat pump sales had doubled between 1975 and 1976 to 312,000 units [66]. By 1978, almost another doubling took place with shipment of 560,000 units [67]. Although 1979 figures are not totally compiled, heat pump sales were roughly even with 1978 [68]. This levelling off occurred due to the announcement of additional natural gas supplies, causing a surge in the purchase of gas furnaces. (See Figure 26.)

Realizing its vulnerability as an oil importer, Western Europe is carrying on much heat pump research. Westinghouse Electric in Belgium has conducted tests of actual air source heat pump installations. They report a Seasonal Performance Factor of 2.3 in Northern European locations and 3.5 in Southern Europe [69]. The Dutch government is investigating heat pumps with direct drive and waste heat recovery [70]. The Swedish Council for Building Research conducted a 1975 study of future heat pump applications within that country [71].

Different European countries are making use of various heat sources in their present heat pump applications [72]. Air source heat pumps are found mainly in central and southern Europe. Ground and water wells are the more common heat sources in northern Europe. Plants which utilize solar/heat pump combinations for heat delivery have been built in some countries. Denmark and Switzerland currently lead the continent in heat pump installations [73].

There is a growing interest in the potential impact which the heat pump will have in the future energy mix of Europe and America, as well as in other countries. On the European scene, market research analysts have predicted that heat pump sales will more than double from $135 million in 1980 to $300 million by 1985 [74]. One-third of all solar installations, they feel, may ultimately include a heat pump [75]. Some experts expect that the heat pump will replace most oil-fired central heating systems in European residences [76].

In the study conducted by the Swedish Council for Building Research, two possible scenarios were outlined for heat pump installations in the year 2000. (See Figure 29.) One alternative considers oil prices as more attractive than electric - hence there is no incentive to install electric heating. The second case assumes that the cost of heating with electric is lower than with oil. Approximately 40% of electrically heated buildings are assumed to use a heat pump in either case. The figures listed might combine to average a total heat pump penetration of 10% for Scenario I and 20% penetration for Scenario II for all building types combined.

Building Type	Scenario I (If Oil is Favoured) %	Scenario II (If Electric is Favoured) %
One-Family	12	36
Apartment	2.4	14.4
Other	14	22

Fig. 27 [77] Percent of Swedish Buildings with Heat Pump Systems - Year 2000

In the United States, the survey of heat pump technology sponsored by the U.S. Department of Energy and prepared by Gordian Associates, Inc (New York) in mid-1978 contains a detailed unitary heat pump market projection.

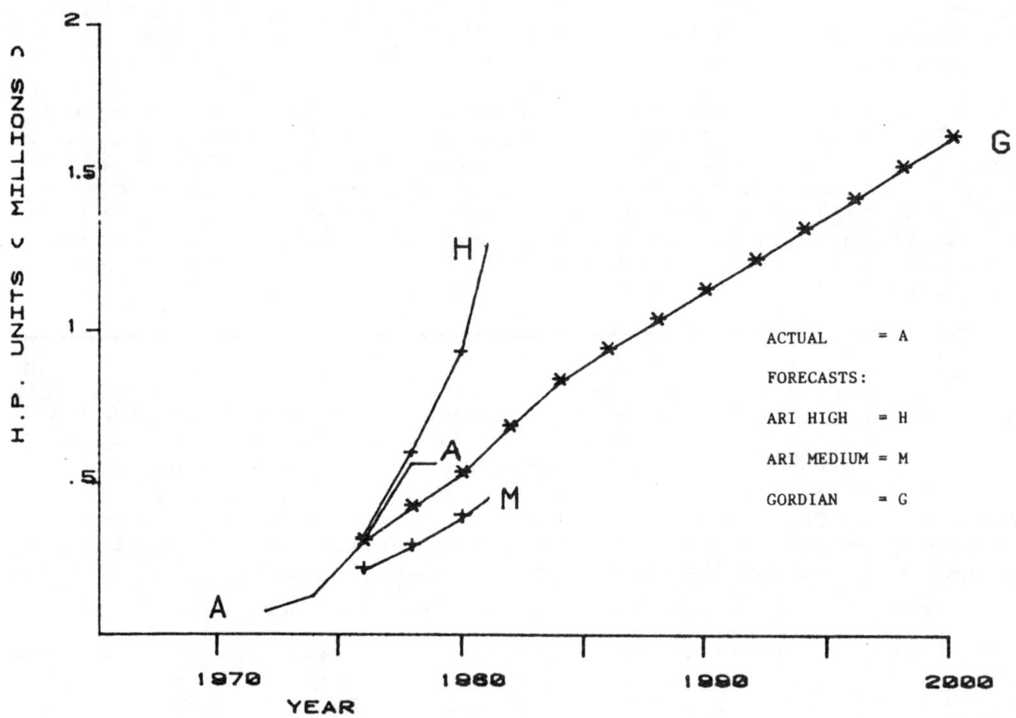

Fig. 28 Standard Elec. H.P. Forecast U.S. Market

Figure 28 depicts the actual United States heat pump shipments from 1972 to 1979 along with forecosts made by the Air Conditioning and Refrigeration Institute and Gordian Associates [78]. The figures from 1976 to 1979 reveal that actual shipments fall between ARI's high estimate and the Gordian prediction. The Gordian projection is based upon manufactuerers' estimates. The heat pump's share of the central heating equipment market rose from 4.7% in 1973 to 8.3% in 1975 [79]. The 312,000 units shipped in 1976 were more than double the amount of sales one year earlier. A continuation of this trend leads Gordian to predict an estimated 46% market penetration rate by 1990 (approximately 1.13 million units) and 63% by 2000 [80] (about 1.625 million units).

Mid-Range Forecast

Year after Commercialization	Advanced Electric H.P. (# of Units)	Gas Heat Pump (# of Units)
1	15,000	12,000
10	150,000	100,000
20	370,000	220,000

Fig. 29 [81] Projected Domestic Market for Prototype Heat Pumps

An effort to assess the market penetration of gas heat pumps and advanced electric heat pumps which are not yet commercially available is treated cautiously in the Gordian report to the Department of Energy. Two possible scenarios are outlined. A high growth scenario follows the growth rate of the electric air conditioner which met good initial market acceptance. The low growth scenario is based on the early marketing history of the heat pump before service and reliability problems were addressed. Figure 29 lists the mid-range of these two scenarios. An assumption is made that the selling price of advanced heat pumps will be no higher than 30% above standard heat pump cost, as expected by manufacturers [82]. Since commercialisation of a new heat pump may take a minimum of seven years from its feasibility stage, sales estimates are based upon the length of time after a concept's arrival on the market [83]. The estimates indicate a healthy potential market for prototype heat pump systems in the first 20 years after commercial availability. Electric heat pumps were projected to advance beyond gas heat pumps because of dwindling natural gas supplies. The relative sales of electric and gas powered heat pumps will depend on the ultimate availability of these fuel sources.

- Swing from electric furnace to heat pump
- Heat pump sales will parallel new construction starts
- Water-source heat pump may outpace air source
- Add-on heat pump with oil furnace now cost competitive
- New gas supplies will slow heat pump sales - only in the short run

Fig. 30 U.S. Heat Pump Predictions

Over the past year, various predictions have been ventured by HVAC company representatives concerning the path of future heat pump installations. The vice president of Weatherking, Inc. foresees a swing away from the electric furnace toward the heat pump with savings of 40% to 50% in winter fuel bills [84]. Future heat pump use will continue to be affected by the amount of new residential construction. Addison Products sees the future of water source heat pumps as outpacing air source heat pump applications [85] (thus affording more stable heat source temperatures in colder climates). A Borg-Warner Corporation representative claims that the dramatic increase in the cost of oil has now made the add-on heat pump with an oil furnace economically viable [86]. Although the emergence of new gas supplies will slow the rate of growth of heat pump sales, this will not diminish the role of the heat pump in the long run, according to the Chairman of the board of the Trane Company [87].

These prognostications are evidence of the rapidly changing role of the heat pump in the heating and cooling industry at the present time. In view of the vast energy problems facing all nations today, no stone must be unturned in our efforts to curtail fuel consumption. No one solution will suffice, but the heat pump has a vital and growing role to play in the years ahead.

REFERENCES

1. ASHRAE Handbook, 1975 Equipment, p. 43.1.

2. ASHRAE Handbook, 1977 Fundamentals, p. 1.6.

3. ASHRAE Handbook, 1977 Fundamentals, pp. 1.2, 1.7, 1.8.

4. "Heat Pump - Limitations and Potential", Comly et al., General Electric, Sept. '75, p.2.

5. "The Heat Pump", T. Pannkoke, P.E., Heating/Piping/Air Conditioning, Feb. '75. p. 23.

6. ASHRAE Handbook, 1976 Systems, p. 11.1.

7. See Ref. 4, pp. 2-6.

8. "An Investigation of Methods to Improve Heat Pump Performance and Reliability in a Northern Climate", Westinghouse for EPRI, Vol. 1, Jan. '77, p. 2.2-1.

9. See Ref. 8, p. 2.2-25.

10. See Ref. 8, p. 2.2-3

11. See Ref. 4, pp. 7-10.

12. "Heat Pump Technology, A Survey of Technical Development Market Prospects", Gordian Associates for USDOE, Jun. '78, pp. 42, 43.

13. See Ref. 12, pp. 56, 57.

14. See Ref. 12, p. 57, 58.

15. See Ref. 12, pp. 105-111.

16. See Ref. 12, pp. 75-95.

17. See Ref. 5, p. 24.

18. See Ref. 12, p. 60.

19. "Phoenix House: Solar Assisted Heat Pump System Evaluation", Jardine et al., Kaman Sciences Corp., CO, for EPRI, Mar. '78.

20. See Ref. 19, p. 3-1.

21. See Ref. 19, pp. 2-34, 2-35.

22. "Solar Energy Heat Pump Systems for Heating and Cooling Buildings", Proceedings of Workshop conducted by Penn State University and USERDA. Paper presented by E.E. Drucker, Syracuse University: "Commercial and Instituttional Building Solar Heat Pump Systems", Jun. '75, p. 63.

23. See Ref. 22, paper presented by R. Romancheck, Pennsylbania Power and Light Co.: "A Solar Assisted Cascaded Heat Pump Concept", p. 121.

24. "Solar Heat Pump Combinations for a Cold Climate", J. Healey et al., Atmospheric Sciences Research Center, Oct. '79, pp. 5-2, 5-4, 5-5.

25. See Ref. 12, pp. 122, 124.

26. See Ref. 24, p. 2-10.

27. See Ref. 22, paper presented by R.C. Barnett: "Performance Characteristics of Air to Air Heat Pumps", p. 184.

28. "Energy Flow Patterns for 1975", R.B. Kidman et al., Los Alamos Scientific Laboratory, U.S. Government Printing Office, 1977, p. 72.

29. See Ref. 12, Chapter 12.

30. See Ref. 12, p. 10.

31. "Coleman Expects Heat Pump to Shine", New York Times, May 1, 1979, pp. D1 and D16.

32. Coleman tech sheet on the "Energy Conserver" kit and T.H.E. Heat Pump, Coleman, Wichita, Kan.

33. "Energy and Your Home", York Heating and Air Conditioning, York, PA, p. 17.

34. "The Carrier Heat Pump Air Conditioner", No. 838-077-0777, Carrier Corp., Syracuse, N.Y.

35. Air Conditioning, Heating and Refrigeration News (AH&RN), advertisement, Sep. 25, 1978, pp. 20, 21.

36. "Through-the-Wall Heat Pump Units", AH&RN, Jun. 11, 1979, p. 17.

37. "York Defends Reliability of its Demand Defrost System", AH&RN, Aug. 13, 1979, pp. 3 and 12.

38. "York Sees Growing Share of Expanding 1980 Market", AH&RN, Sep. 10, 1979, pp.1 and 18.

39. "Honeywell Will Install Two York Liquid Chillers in New Solar System", AH&RN, Dec. 5, 1977, p. 27.

40. "B-W's New Inverter Drive Applied to Centrifugals, Home Heat Pump", AH&RN, Oct. 8, 1979, pp. 1, 3 and 30.

41. "Exxon Device Claims 25% Cut in Cooling Energy Use", AH&RN, Jun. 18, 1979, p. 18.

42. "Defrost, the Heat Pump's Nemesis - a New Solution?", AH&RN, Jun. 5, 1978, p. 1.

43. "Process Heat to be Supplied by Westinghouse Heat Pumps", AH&RN, Jun. 5, 1978, p. 26.

44. "Sawing Solar Costs in Half", A. Weinstein et al., Westinghouse Electric Corp., Solar Heating and Cooling Div., 1978.

45. Advanced Energy Technology, Clifton Park, NY News Release, May 14, 1979.

46. "Senate Panel OK's Tax Break on Heat Pump", AH&RN, Oct. 1, 1979, p. 1.

47. "DOE Awards 7 Proposers Heat Pump Contracts", AH&RN, Mar. 19, 1979, p. 31.

48. "TVA Loans are Available for Energy Saving Measures, Including Heat Pumps", AH&RN, Sep. 10, 1979, p. 37.

49. "LILCO Sponsors 2-Year Energy Storage Demonstration Program", AH&RN, Jul. 9, 1979, pp. 3, 5.

50. See Ref. 24, Table 5-2, p. 5-4.

51. See Ref. 12, Exhibit 9-4, p. 228.

52. See Ref. 12, pp. 222, 228, 265.

53. "Despite Winter Conditions Heat Pumps Perform Well", AH&RN, Apr. 18, 1977.

54. "Elements of Meteorology", Miller and Thompson, p. 214.

55. "Weather Data and Design Conditions", ASHRAE Fundamentals, Chapter 23, 1977.

56. "Swedish Alternating Ventilation System...", AH&RN, Oct. 1, 1979, p. 16.

57. "U.S. Agency Issues New Energy Forecasts for '85, '90, '95", AH&RN, Jun. 28, 1979, pp. 1, 33.

58. "Icy River Helps Heat Home", AH&RN, Sep. 3, 1979, p. 3.

59. "Home Builders' Association Sees Recession in 80's", Schenectady Gazette, Dec. 7, 1979.

60. See Ref. 12, p. 302.

61. See Ref. 12, p. 308.

62. "Europeans Continue to Set Pace in District Heating Applications", AH&RN, Sep. 17, 1979, p. 9.

63. "Does the United States Waste Energy?", EPRI Journal, Nov. '79, pp. 26-32.

64. "Energy-Short Europeans Bunds Up for Winter, Seek Fuel Saving Measures", Schenectady Gazette, Nov. 8, 1979, p. 16.

65. See Ref. 12, Exhibit 10-1, p. 272.

66. See Ref. 12, p. 245.

67. See Ref. 38, p. 18.

68. "Replacement Air Conditioning Business Gains Market Share Increased Attention", AH&RN, Jul. 2, 1979, p. 6.

69. Total Energy: Proceedings of the First International Total Energy Congress, Copenhagen, Denmark, Oct. 4-8, 1976, pp. 880, 881.

70. See Ref. 69, p. 878.

71. Heat Pump Applications in Sweden: Research and Development Needs, Zahir Fikri, Swedish Council for Building Research, 1975.

72. Solar Energy Systems and Components Market in Western Europe, Frost and Sullivan, Report No. E251, New York, NY, 1978, p. 37-II-26.

73. See Ref. 72, p. 359-VI-130.

74. "Solar Energy to Contribute 2 Percent to European Energy Consumption by Year 2000", Solar Heating and Cooling, Gordon Publishers, Morristown, NJ, Oct. '79, p. 19.

75. See Ref. 72, p. 360-VI-131.

76. See Ref. 72, p. 359-VI-130.

77. See Ref. 71, p. 41.

78. See Ref. 12, p. 278 for ARI and Gordian forecasts and 1970-76 actual and Ref. 38 for 1978-79 actual figures.

79. See Ref. 12, p. 245.

80. See Ref. 12, p. 258.

81. See Ref. 12, Exhibits 10-9, 10-12, pp. 280, 283.

82. See Ref. 12, p. 257.

83. See Ref. 12, p. 268.

84. "Weatherking Introduces New Products, Services for '78", AH&RN, Jun. 5, 1978, p. 13.

85. See Ref. 68, p. 6.

86. See Ref. 38, p. 18.

87. "Opportunities Still Abound in Less-Than-Robust Economy", AH&RN, Nov. 29, 1979, p. 3.

THE LOW ENERGY ADAPTABLE DWELLING (LEAD) PROJECT*

D. Forrest

Autarkic Associates, 24 Charlotte Street, London W1P 1HJ, UK

INTRODUCTION

Essentially, the design is for a low energy house. Energy and resource use has been considered in three principal ways. They are:

1. In the resources used in the construction of the structure and envelope.

2. In the performance of the dwelling in energy use by energy conservation.

3. By ambient energy collection.

Besides the matter of the resources used in its construction initially, the dwelling design also seeks to cater for the over-lays of rapidly advancing technologies, outside the energy field, and certain other social factors, which may eventually affect family life styles, and therefore demand greater changes in facilities and spatial arrangement within the house. The structure is designed to have a greater degree of flexibility in its internal planning than many current house types. This is one reason for the use of a post and beam structure. Such a system encourages more adaptation and may even be more economic than the panel system initially also.

The design concept allows for both the energy saving techniques which are cost-effective at this time and also for others which may be incorporated later as 'customer' options. These will become viable as technology advances and should therefore be properly allowed now. Therefore the dwelling aims at being commercially viable in the private and public house building sectors now. It also allows and anticipates the future. These may certainly be considered shortly as values, both as installation and running costs themselves change.

The objective at this time, is to demonstrate the fulfilment of these objectives. Our opinion is that by practical demonstration with properly monitored results the benefits of such dwellings may be convincingly shown to the industry and the public alike.

---

* See Editor's Note at the end of paper.

The motivation for most people to consider employing ways of saving energy, (the choice of materials and design flexibility) are almost always those of simple economics. These arguments may not always be thought to be important to the majority of people at this time but will almost certainly be so in the future. As a guess, more general appreciation of energy economy in the home as an important factor may be forced upon many families within the next four to five years.

The object of this house unit concept is to appeal to the general, rather than the more progressive sector of the British public.

CONCEPT

As a starting point for the design, the volume of the dwelling has been considered as a cube.

In any design there are a number of relatively permanent features in dwellings (i.e. the first floor platform, the stairs and the envelope) but adjustment to these features may be more conveniently considered within a cube either at the initial construction stage or say after a ten year life. (This is to acknowledge that the frame is also designed to allow that degree of flexibility.)

The photograph of the completed frame shows the structural implications of the system. However, it does not at this stage show any developed planning or the visual implications. It is only the basis upon which to develop the design and to complete the actual structure.

The design may be contracted (by overlapping to provide say 2 bedroom houses in semi-detached or terraced form) at the initial building stage.

If the structure is completed in cubic form, that shape inherently allows a greater degree of flexibility in five directions (sideways and upwards). It also allows for the minimum practical exterior envelope surface. Therefore in both the use of materials and protection from the elements, a cube has some advantages in heat energy conservation. In planning also, the basic cube with a central vertical circulation volume, allows a considerable degree of alternative planning forms, and perhaps equally importantly, practical alternative volumetric arrangements in section. The cube is less inhibiting than many other shapes in this respect.

The design includes one less conventional house feature in the U.K., a semi-basement. This feature besides being able to act to conserve energy in its own right, also allows additional storage. The storage facility besides catering for today's needs, also allows for those now required by energy considerations. These may include provision for heat energy storage (short term now, longer term in the future) certain items of plant (for example, heat pump equipment, additional trunking etc). The roof space, previously considered as a storage facility may, through increased insulation and through greater demand for plant, principally ventilation, be less available as certain techniques for ventilation and solar gain are included.

FRAME

In considering the frame, besides the question of flexibility and economy of materials, ease of erection has been considered also. Such will facilitate erection on site by unskilled labour or even, if required, allow D.I.Y. involvement. The photographs shown here indicate this latter method of construction. The "meccano" parts and their joints may be designed to ensure that the correct sequential procedure is followed.

At this time we believe that certain scantling sizes may be decreased if that is required. However, at this stage, such economies may not be warranted.

The diagrams have been developed to show various internal arrangements, principally for stairs and floor panels. We suggest that the central stair solution is accepted at this time. It has some inherent advantage given certain conditions.

The photographs show a perimeter constructed as a simple post and beam construction. The purposes of the perimeter frames are two: to allow for ease of erection and, if required, to facilitate increased flexibility to "stretch" or change the envelope at any later stage. However, if as discussed later, it is agreed that a solid internal lining has certain energy advantages, that perimeter frame may also be omitted to economise on cost if that is considered of prime importance.

The central crusiform columns from which the bi-partite structure is central to the design of the frame. The purposes of these twinned beams, which logically follow, may be summarised as encouraging flexibility, allowing small service voids (important features in energy design) and to also allow resource economy by using smaller scantling sizes and therefore also to increase the availability of timber. The pattern created by this system within the dwelling expresses some purpose and discipline also, an important visual aspect if one is to consider later adaptation on a D.I.Y. basis.

Jointing has been considered in detail. As far as is possible, plated or "connected" junctions will be avoided. More traditional methods will be considered.

The pyramidical roof allows that part of the structure to fulfil two functions, a structural one and to better resolve the question of aspect. The ring beam (in effect a "notched" plate) performs a tie function acting in unison with the rafters. The four aspects of the roof over-come certain siting problems. In ambient energy collection, and more particularly in solar energy, such aspect consideration is important. The roof has been designed to allow such solar collection features to be easily incorporated, its form imposes less limitation in site design than a roof which faces towards two, rather than four, points of the compass.

CLADDING

If, given the choice, the majority of house purchasers appear to prefer houses, which if the structure is not constructed of mostly "permanent" materials, has the overall appearance of being so constructed, the cladding of a timber framed building is important. The proportions of say brick to other claddings (timber, glass etc) generally acceptable are difficult to establish precisely. Many factors play a part. This must be properly considered as the design is finally developed.

In many respects, in the consideration of resource and energy use, the matter of external appearance may present a conflict of design priorities. This is especially true of the use of glass, where if one is to consider passive solar collection, large areas of glazing may need to be employed in the design. Through our calculations, checked by computer, we have established our correctness of employing such areas of glass. The degree of so doing may become a more subjective judgement, but the benefits of so doing will also enter into that judgement.

And, as has been previously mentioned, mass on the internal lining has some quite definite advantages in energy use terms. In fact this may be achieved by relatively "dry" methods of construction without resource to special blocks. This being so, there are also other advantages on site. This matter of the employment of solid rather than lightweight internal walls may be a relatively unique feature of the design at this time.

The alternative and more commonly constructed skin at this time is lightly constructed, but highly insulated panels. However, the disadvantages of low thermal mass may actually be an advantage in certain living (and working) volumes and, given certain conditions, may also be shown to be effective. Certainly such panels have many advantages and, as such, should also be included in the final house unit, but essentially the house has a thermally controlled mass.

The prime considerations, in energy terms, of this exercise may be summarised as:

1. Improved heat energy consideration by simple conservation.

2. Improved heat energy conservation further improved by limited and simple ambient energy collection.

3. Further increased performance through passive solar gain, all buildings inherently being solar collectors, albeit mostly uncontrolled.

4. Various refinements to improve a performance of individual buildings.

The first two objectives, (if not already demonstrated by the housing types of the developer) may be so shown by using their catalogue of existing house types as further demonstration of these features on this site. The second two objectives may be shown through the "LEAD" house types, commencing essentially with the collection and optimisation of solar gain through more passive means.

EDITOR'S NOTE

Although the LEAD project is not primarily concerned with solar energy, it nevertheless suggests practical ways for the integration of solar solutions in energy conserving dwellings. Mr. Forrest's contribution in these conference proceedings is of particular relevance in view of the realistic directions which it points to for the development of solar energy solutions in the building field. The Low Energy Adaptable Dwelling concept, as it is described in this paper, offers vast possibilities for the use of both active and passive solar systems. At this stage it is far more important to show the existence and potential of such possiblities than to detail the actual solutions. The flexibility of the LEAD project offers a great variety of solar solutions.

Model Showing the LEAD Building Under Construction

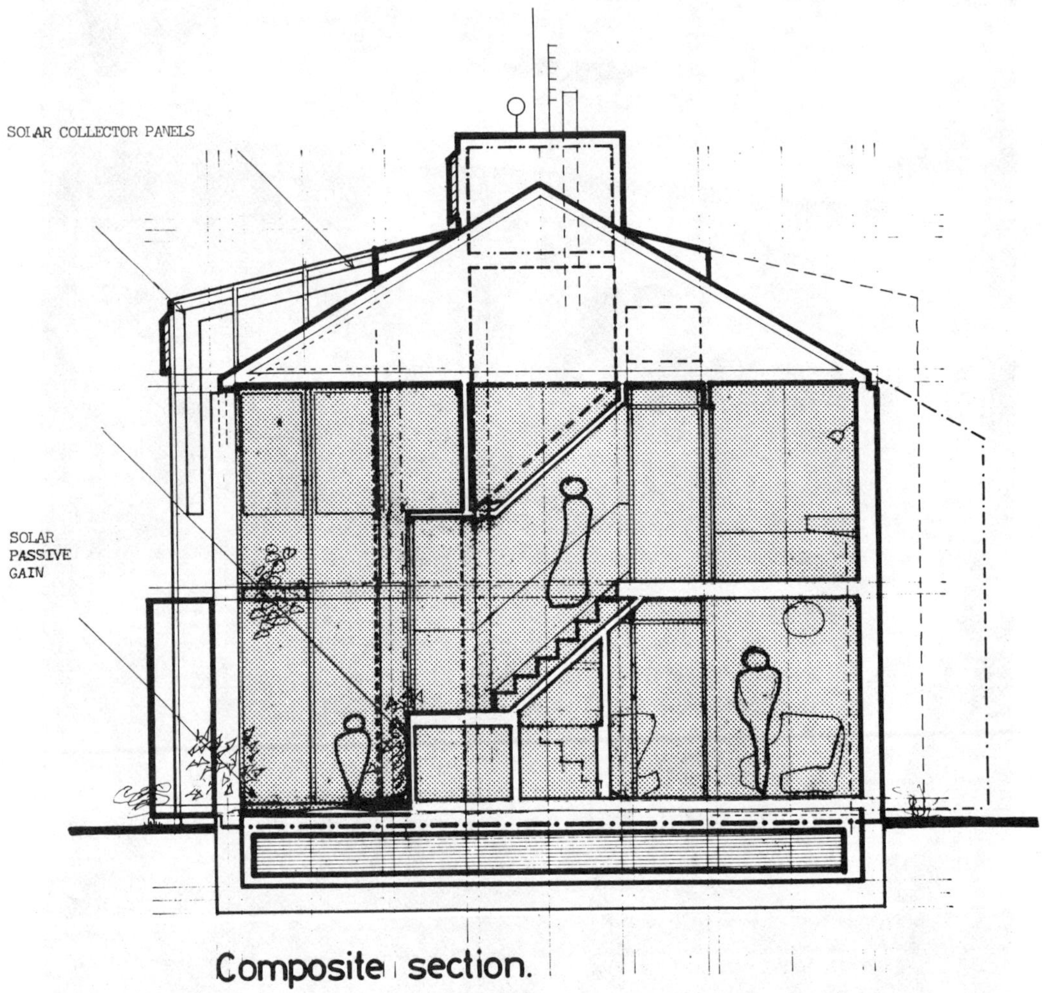

Composite Section

EVERY MAN'S SOLAR ENERGY

A. Russell-Cowan

Solar Centre, 1A Fane Street, London W.14, U.K.

ABSTRACT

The paper discusses the problems which are inhibiting the wide diffusion of solar collector technology. The task of increasing collector efficiency regardless of cost implications is identified as one of the most limiting factors. Several examples are described of low cost collectors and finally various important design and installation problems are identified which require further research.

KEYWORDS

Efficiency; cost; low cost collectors; thermosyphon; trace-coil; low profile; combine collector; tank.

1. INTRODUCTION

Most of the known technology and practical application of harnessing solar energy has taken place in the last 80 years and especially during the past 8 years there has been an ever increasing public interest and a concentrated effort by scientists and engineers to make improved solar products. Most of the efforts have been directed towards flat plate collectors for heating water and a study of solar collectors made forty years ago clearly indicates that we have made little progress in this field. Basically in only two ways, namely the inclusion of better insulation and the application of black selective surfaces.

All recent endeavours have resulted in the development of highly expensive products that are only available to a few 'well-off' people, mostly living in the northern industrial nations. Countries such as USA, UK, France and Germany who now control the market are pricing their products out of reach of most people even in their own countries but particularly in the developing world. The reason given is inflation, the ever increasing cost of materials and labour. Also that a 'cottage' industry cannot obtain sufficient orders to turn from 'batch' to 'mass' production. The real answer surely is that most solar products are over-designed and that the attainment of a little more efficiency at higher prices is considered the main criterion.

We also have the 'wait and see' attitude of the great companies who are suspicious of an effective simple technology and are only prepared to carry out a 'watching brief'.This, inspite of the fact that many of these companies are now benefitting in additional sales of their products that are used in solar systems, such as metal, plastics, pumps, electronic controls, insulation piping and glass etc.

Lastly, there is a sceptical attitude of governments who are mainly only prepared to place small amounts of money into research that is being duplicated over and over again and are not prepared to subsidise manufacture or offer tax relief or sufficient benefits to those who save energy. We are therefore guilty of restricting the sale of our products on the home market and almost completely preventing them from being widely available to the most suitable sun-rich countries of the world whose development is being retarded by a shortage of power supplies and therefore need expensive hand-outs for survival.

We are guilty of ignoring the export potential of our solar products and 'know-how' to perpetuate this new industry and for the future there must be a new approach in the early 80's to take solar energy to the people. More effort should be directed in the design and use of new materials for simple proven applications with an emphasis on low cost and not just improved efficiency. The issue is: The most BTU's for the least amount of money. It is appropriate to quote here the late Farrington Daniels who wrote:

> "To make solar energy competitive with fossil fuels universally depends entirely on the development of cheap practical solar collectors." (1)

2. SOME LOW COST SOLUTIONS

We should look again at the oldest and most widely used system for collecting solar energy, the thermosyphon or natural circulation method, and improve it for even greater use. It still represents the most cost effective utilisation of solar energy. Even in areas of moderate radiation, a thermosyphon system can be used for the greater part of the year. Different types of thermosyphons systems have been developed and some of them are described as follows:

Trace Coil:

Direct heat transfer to the outer surface of storage cylinders from solar collectors (see Fig. 1).

Pipe tracing systems are widely used in industry and carry steam, hot water, hot glycol and other suitable heating fluids.

A cheap method of using solar energy is suggested by the use of a trace coil which is easily fixed to the exterior wall of an uninsulated storage cylinder, greatly reducing plumbing installation costs and the eventual risk of anti-freeze mixing with the hot water if a leak occurs in the heat exchanger coil. Trace coils can be fixed to direct cylinders. The trace coil in the use proposed is in effect an extension of the solar collector plate or an external heat exchanger.

As in all solar products, thermal contact is of vital importance and this is achieved in trace coils by first clamping the coil with stainless steel bands around the perimeter of the existing hot water cylinder and then sealing air gaps with a heat transfer cement, e.g. 'Thermon'. The non-hardening heat transfer cement

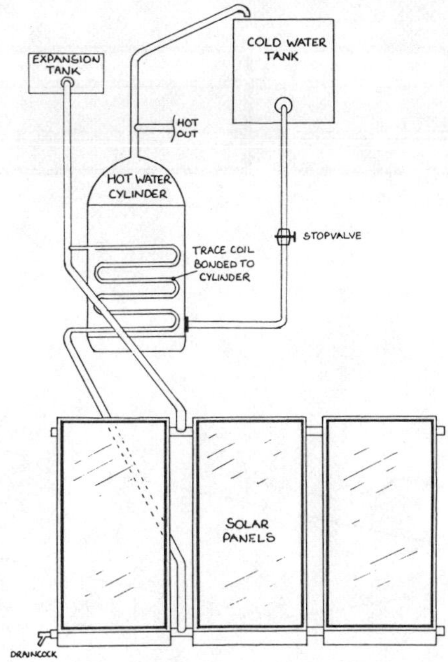

Fig. 1 Thermosyphon System Using Trace Coil

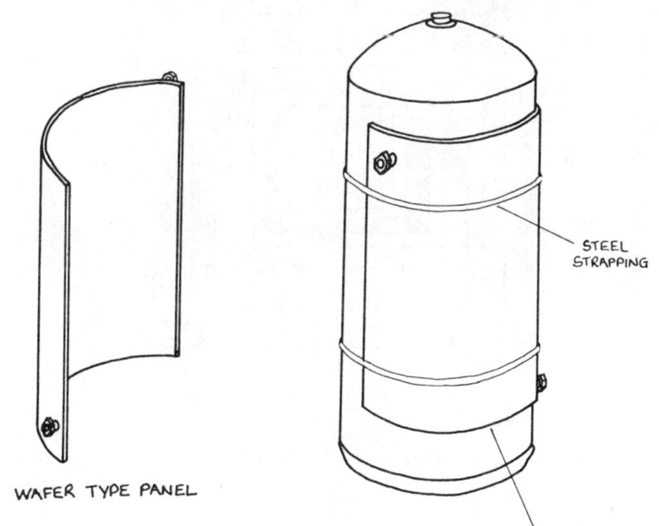

Fig. 2

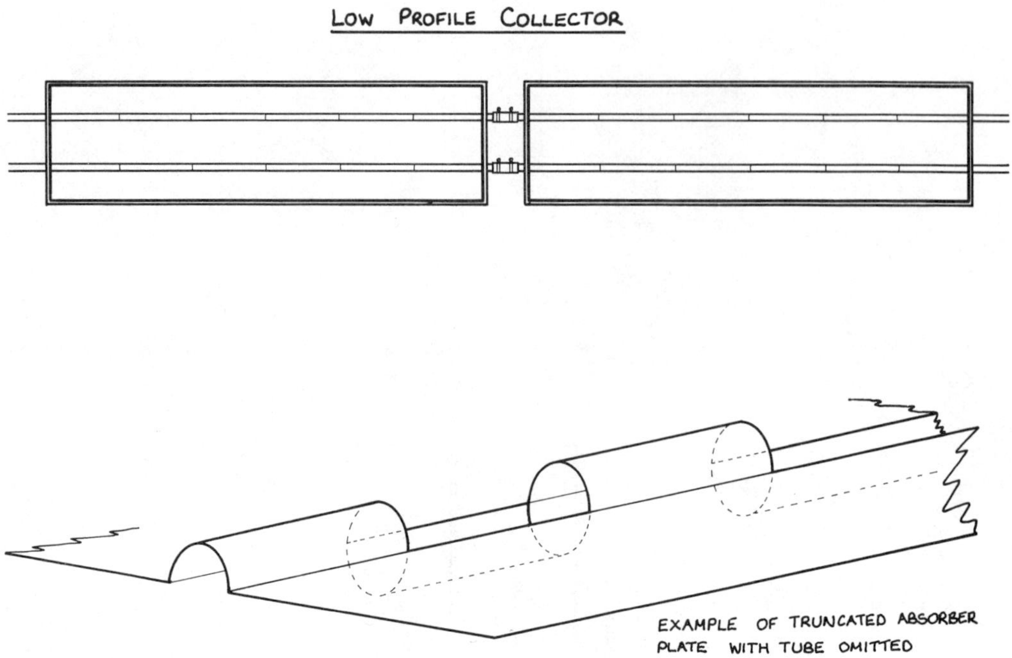

Fig. 3

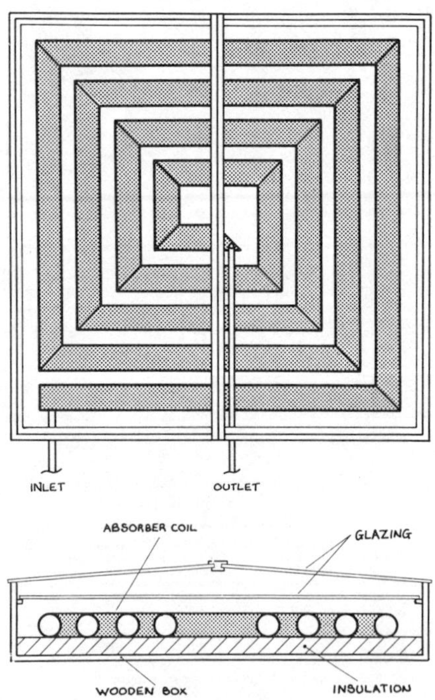

Fig. 4 Combine Collector by Chinnappa

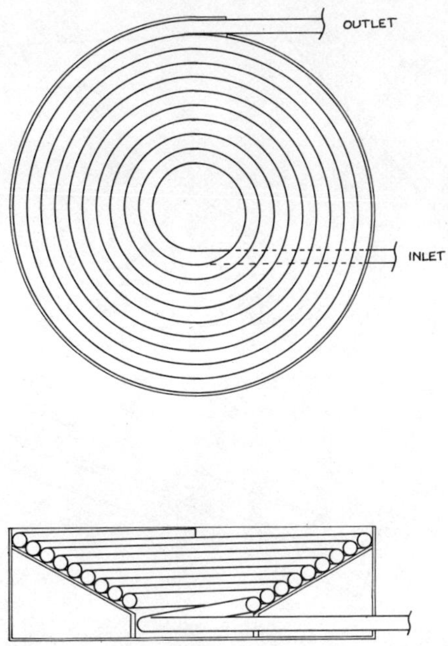

Fig. 5 Thermosyphon Coil

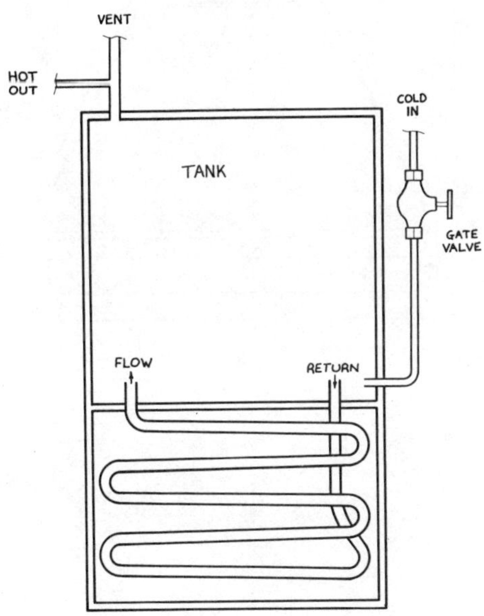

Fig. 6 Tank With Integral Collector

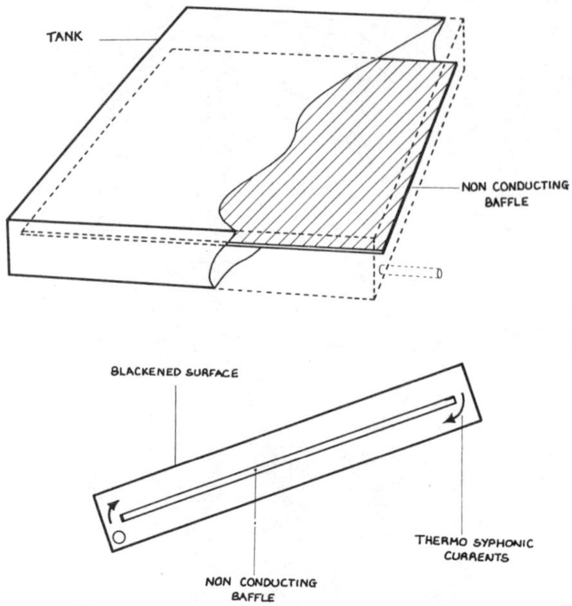

Fig. 7 Tank With Internal Baffle Plate

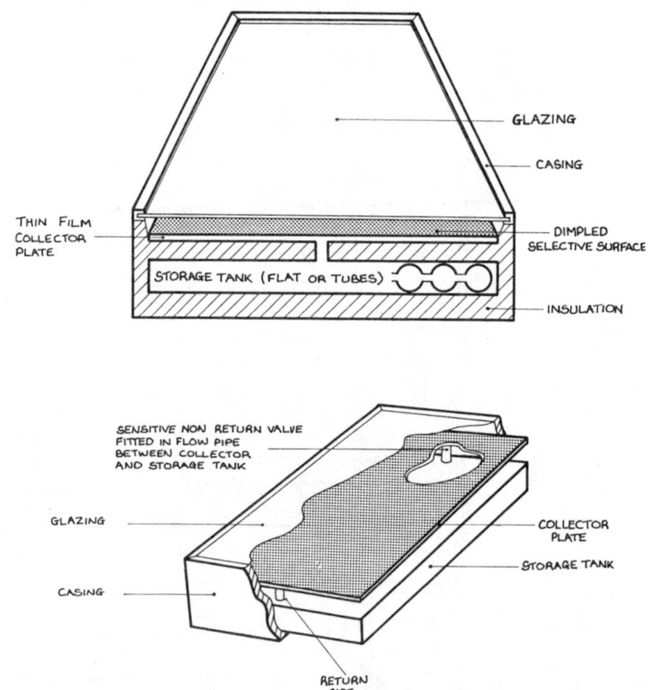

Fig. 8 Cross Section Through Proposed Combine Collector

Tank with Integral Collector

Many successful combine collectors have been made by the manufacturers of a galvanised rectangular tank varying in size from 8 ft^2 (.74m^2) to 20 ft^2 (1.86 m^2) and in depth from 2 to 4" (50.8 - 101.6mm). The front of the tank is blackened to see the sun - the water being in contact with the blackened surface. The whole unit is set in an insulated glazed box and tilted to the correct inclination.

Fig. 6 shows an improvement to the ordinary flat tank by the addition of the extension of the tank's top surface to which is fixed a serpentine coil which improves collection of energy by a thermosyphon action.

Fig. 7 is a further improvement to a tank collector by the inclusion of a non-conducting baffle or septum set near the top surface of the tank permitting only a small amount of water to be heated at the front thus accelerating a thermo-syphon flow.

3. IMPROVING THERMOSYPHON SYSTEMS

A number of ultra-sensitive non-return valves have recently been invented in various countries which inhibit the flow of water into the storage tank until a predetermined temperature is reached (6) (7). A second function of this device eliminates the necessity of having to place the tank above the collector, enabling the storage tank to be placed anywhere in the building. These non-return valves, used in conjunction with combine collectors can greatly enhance the use of thermo-syphon systems.

A proposed combine collector (see Fig. 8) indicates the use of the non-return value in combination with a thin-wafer type collector with a maximum area of contact between the flowing water and the metal surface exposed to the sun which results in more uniform temperatures across the metal plate.

A new form of combine collector has been manufactured based on the use of standard mass produced copper storage cylinders with half the cylinder surface blackened and facing the sun. Within the cylinder is a semi-circular non-conducting baffle or septum which is insulated from the main body of water resulting in an internal thermosyphon action. The back half of the cylinder is insulated (see Fig. 9).

An extension to the base of the cylinder in the form of a curved thin wafer causes a further small amount of water to heat rapidly. There is therefore a double thermosyphon action. Recent tests in Spain in winter produced excellent results and it is hoped that the product will shortly be marketed.

A cyclindrical collector receives more radiation than a flat plate collector of the same proportions by offering a more constant surface to the sun's diffuse rays during the day. Reflective losses from a curved surface are less than from a flat one.

Most of the research and commercial work carried out during the last four years on solar energy systems has been directed towards pump control systems which are still not fail safe and are more costly to manufacture and install. Many researchers have confirmed in dual experiments that there is no appreciable difference in collector efficiencies between pumped and natural circulation systems. Per haps now is the time for the solar scientists and engineers to have another look at thermosyphon systems and resolve some of the outstanding problems

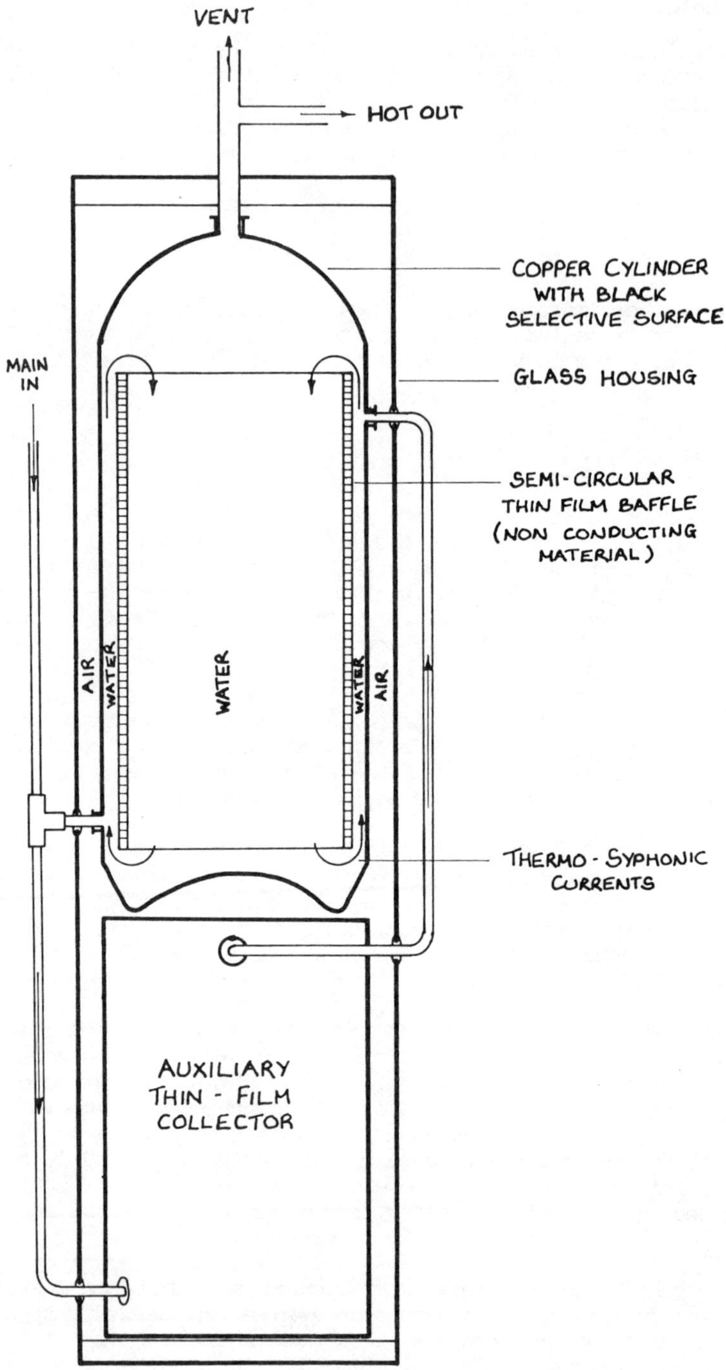

Fig. 9 Cylindrical Solar Collector

which are listed as follows:

1. Seek methods of precipitating water flow in relation to a reduced head.

2. Incorporate the new selective surfaces.

3. Incorporate storage tanks in the interior of the collector boxes which would result in a preheated storage adjacent to the collector with minor heat losses at night due to short pipe runs, non-return valve and selective surfaces.

4. Investigate all the parameters to find ways of making natural circulation systems more effective with smaller areas of collectors.

5. Develop an automatic drain down system in order that thermosyphon systems could be safely used in areas of moderate radiation and freezing.

6. Improve the new sensitive non-return valve and reduce the cost.

7. Investigate new materials that permit freezing without damage by coupling metals with plastics that stretch.

8. Investigate the optimum depth of combine solar storage tanks.

REFERENCES

(1) Farrington Daniels, 'Direct Use of the Sun's Energy', Yale University Press, 1964.

(2) J.C.V. Chinappa and K. Gnanalingam, 'Performance at Colombo, Ceylon, of a pressurised solar water heater of the combined collector and storage type', Solar Energy, Vol. 15, pp. 195-204.

(3) H.P. Garg, 'Year round performance studies on a built-in storage type solar water heater at Jodhpur, India', Solar Energy, Vol. 17, pp. 167-172.

(4) A. Russell-Cowan and C. Stambolis, 'Development of built-in storage type solar water heaters for large scale use in developing countries', in Solar Technology for Building, Ed. C. Stambolis, RIBA Publications Ltd, London 1978.

(5) F.L. Thannhauser, U.S.A., Patent No. 3254644, June 1966.

(6) Private communication to Mr. J.C. Kapur, Kapur Solar Farms, N. Delhi (Oct. 1978).

(7) S. Bucley, Mecahnix Illustrated, December 1976, p. 31.

CONSIDERATIONS FOR ESTABLISHING THE DURABILITY OF SOLAR ABSORBER SURFACE COATINGS

M. G. Hutchins

Solar Energy Unit, University College, Newport Road, Cardiff CF2 1TA, UK

ABSTRACT

A review is presented of degradation mechanisms pertinent to the assessment of the lifetime of solar absorber coatings. Temperature effects, ultraviolet irradiation, humidity, vacuum stability, and compatibility of the coating/substrate interface are discussed. An outline of appropriate durability tests is presented together with results obtained for black chrome and other candidate absorber films.

KEYWORDS

Solar absorber coatings, selective surfaces, degradation, durability testing, accelerated testing.

INTRODUCTION

It is almost inevitable that a device designed for the terrestrial collection of solar energy will spend its working life in an atmospheric environment. The effect of exposure to the hostile conditions of the natural weather cycles will be a gradual erosion and degradation of the integrity of the materials used for the construction of the collector. Efficient, economic design and operation of a solar collector is a trade-off between a set of materials which possess favourable optical and thermal properties and a choice of compatible materials which can be relied upon to provide a relatively long, trouble-free lifetime involving a minimum of maintenance. Many solar energy applications, particularly in more northerly locations, suffer from the economic disadvantage of requiring a full-scale auxiliary back-up system such as a conventional gas boiler which will be called upon to provide the demand load when an insufficient quantity of solar radiation is available e.g. in winter. Thus initial high capital costs can only be off-set against real savings in fuel costs. At current market prices this often makes solar energy systems appear 'uneconomic' and results in lengthy 'payback' periods. As an example a typical domestic solar water heating system operating in the UK is likely to have a 'payback' period of between 15 and 25 years (Hutchins and co-workers, 1979). This is of course a prediction containing many assumptions and factors e.g. interest rates and fuel prices which may change the situation in future years. Nevertheless in order to ensure a maximum return on the initial investment, and to gain popular acceptance and establish a viable product for the consumer market, a considerable body of effort and study must now be directed towards the assessment of materials performance in solar energy systems with the aim of producing cheaper, more durable and reliable, thermally efficient solar systems.

To date only a minimum of data exists on materials performance which allows some degree of informed evaluation of the relative trade-off between production costs and predicted lifetime. In addition the accelerated testing of materials and the correlation of materials performance relative to in-service conditions have not been fully established. It is within this context that the following considerations outlined in this report have been made.

It is the aim of absorber surfaces R & D programmes to produce cheap, durable, cost-effective coatings and to contribute to the overall improved efficiency of solar system performance. In order to fully assess the advantages to be gained from improvements made in surface coatings it is strictly necessary to undertake a benefit analysis and determine the relative levels of funding in related areas of solar energy R & D, such as improving heat exchangers, heat transfer fluids, reflector surfaces etc., in terms of the decrease in cost of eventual delivered energy. Without such an analysis it can only be said that the successful operation of any solar system relies very much upon the continued operation and reliability of the surface coating and that the use of selective surfaces in solar systems has been shown to improve the thermal efficiency of solar heating systems significantly (Hutchins, 1979a).

SOLAR ABSORBER COATINGS AND MATERIALS PERFORMANCE

The receiver surface is common to all solar thermal systems. Since the postulate of using 'selective surfaces' to achieve higher thermal efficiency in solar collectors (Tabor, 1956) solar absorber surface coatings have been developed for operation over an extensive range of temperatures. Electrodeposited black chrome has appeared as the most popular and successful of a variety of coatings produced by electrodeposition and chemical conversion techniques for operation in flat-plate collectors up to 200°C (Hutchins, 1979b). Sputtered metal cermet and metal carbide films (Sikkens and Francken, 1979) and silicon-stabilised multi-layers (Seraphin and Meinel, 1976) fabricated by chemical vapour deposition (CVD) have been developed for operation in systems employing single and two-axis tracking with surface temperatures up to 700°C (see Figures 1 and 2). For temperatures in excess of 700°C, which are experienced in central receiver systems (CRS) cavity boiler designs, refractory non-selective paints and oxidised stainless steel coatings have been used (Call, 1979) (see Figure 3).

The first test for any candidate solar absorber coating is the measurement of the surface optical properties, solar absorptance and thermal emittance (Hutchins, 1979c) which should be fully characterised over the spectral range 0.25-50μm. Another desirable property of the coating is a high thermal conductance. Also of interest is the response of the coating to variations in the angle of incidence of the radiation involved e.g. solar and thermal. However, it is also very important to be able to determine the coating durability in order to be able to safely predict an operational lifetime for any particular surface e.g. black chrome, and to assist in understanding the mechanisms and processes responsible for the degradation of surfaces.

However, before discussing degradation mechanisms and methods of testing it is worth remembering that the absorber coating should not be considered in isolation. Successful solar energy systems require subsystems which are durable and maintainable. Material problems must be considered in relation to all components used

Fig. 1. Single-axis Tracking Parabolic Trough Concentrator Employing Linear Receiver Coated with Electrodeposited Black Chrome for Operation to 300°C (Sandia Labs, New Mexico, USA).

Fig. 2. Two-axis Tracking Parabolic Dish concentrator with Cavity Boiler for Operation up to 500°C (University of Houston, Texas, USA).

Fig. 3. Central Receiver System (CRS) Solar Thermal Test Facility Capable of Producing Surface Temperatures in Excess of 1000°C (Sandia Labs., New Mexico, USA).

e.g. transparent covers, substrates, insulation, casing,sealants, piping, heat exchangers, pumps and storage vessels, to mention but a few. The integrity of a solar collector or a total system will be affected by the compatibility of the materials chosen to do the job. Failure to take account of the compatibility of materials can manifest itself in many undesirable effects, e.g. (1) galvanic corrosion in the working fluid conduits caused by inadequate inhibitor protection in mixed metal systems, (2) ingress of atmospheric components into the collector caused by failure of seals or inadequate venting arrangements (condensation), (3) undue stressing of covers or absorber plate due to differential thermal expansion of a component and its fixing points often resulting in cracking or warpage, (4) a choice of insulation material that is incapable of with-standing the stagnation temperatures of the blackened absorber plate and the resultant release of vapours which can cause damage to cover and absorber in addition to no longer serving as an adequate heat loss barrier, (5) the loss of mechanical strength in flexible hoses used to join collectors together, (6) the use of aluminium and mild steel in direct systems which continually draw in raw, oxygenated mains water resulting in scaling of tanks and pipe-runs and corrosion, (7) the use of low melt-point solders incapable of withstanding absorber plate stagnation temperatures leading to melting and degradation of the thermal bond between fin and tube.

There are many other examples of the mis-match of materials but it is sufficient here to simply stress that any system should be designed as a whole and care should be taken to satisfy the minimum requirements that will allow the solar system the chance to repay some of its capital cost by functioning adequately for a satisfactory period of time.

ABSORBER COATING DEGRADATION MECHANISMS

Temperature Stability

The solar absorber coating must be capable of withstanding several different situations in which the danger to stability arises from temperature effects. The maximum stable operating temperature of a coating must be determined and compared to the stagnation temperature that would be experienced by the coating if it were under no-flow conditions in a solar collector. For selective surfaces with favourable optical properties employed in flat-plate systems this can be as high as 200°C, and 150°C for non-selective coatings such as paints. The deleterious effect of elevated temperatures on the thermal stability of an absorber surface has been studied (Pettitt and Sowell, 1979). They investigated the effects of employing electro-deposited black chrome selective surfaces as the absorber coatings for linear focus parabolic troughs which operated for more than 3600 hours with working fluid temperatures of up to 310°C. Changes in the optical properties of the coating were observed. The solar absorptance was found to decrease by ~8% from the initial value and thermal emittance rose by a similar amount. These changes in the optical properties were shown to be directly caused by the thermal oxidation of metallic chromium particles which in black chrome are present within a host dielectric matrix of Cr_2O_3 (Ignatiev, O'Neil and Zajac, 1979). The inhomogeneity of the surface film was decreased thus lowering the solar absorptance and the film thickness was increased with a subsequent rise in the thermal emittance. The knowledge gained from the work of Pettitt and Sowell has enabled a black chrome surface to be developed with a larger average metallic chromium particle size which has improved the thermal stability of the coating.

In everyday operation the solar collector and the absorber coating will be subject to continuous thermal cycling which can result in loss of adhesion of the coating to its substrate and changes in optical properties caused by oxidation and alterations of surface structure. The absorber coating must also withstand stresses resulting from temperature gradients in the absorber plate caused by fin design, and the thermal shock of transients i.e. the ingress of very hot water into a 'cold' collector or very low night-time ambient temperatures in winter.

Ultra-violet Irradiation

High energy solar photons present in the ultra-violet component of the incident solar radiation can cause extensive damage to absorber coatings and plastic collector covers. U-V irradiation can lead to oxygen depletion of lattice surface sites, diffusion and induced bound-state defects. This can result in colour changes and an increase in the imaginary part (the extinction coefficient, k) of the refractive index of the material.

Glass possesses excellent weatherability and is not subject to observable U-V degradation. In plastic materials however such as acrylics, polyvinylchlorides, and polycarbonates, free radicals are formed as a result of the removal of hydrogen atoms due to photochemical dissociation. Similar reactions in fluropolymers, polypropylene and GRP, can lead to a loss in tensile strength and subsequent embrittlement and discolouration (yellowing) and a reduction in solar transmittance. Plastic materials can be uv-stabilised by the introduction of uv-absorbing screening agents or excited-state quenchers and these will also add protection to the absorber coating housed under the glazing. The effect of uv-irradiation on solar absorber coating optical stability has received little attention to date.

Humidity and atmospheric effects

The absorber surface is an integral part of the solar collector which as a whole undergoes thermal cycling generally in the presence of changable atmospheric conditions. The combined effects of humidity and high temperatures have been shown to be deleterious to the electrodeposited black nickel selective surface (Peterson and Ramsay, 1975).

The flat-plate collector design is conducive to the formation of condensed water vapour, usually visible on the underside of the transparent cover, and often accentuated by inadequate provision or design of a collector venting facility. In addition to decreasing the solar transmittance of the cover the conditions place an extra stress on the absorber coating.

The local environment of the collector can result in severe conditions. Coastal sites expose the coating to sea air and salt, industrial and urban localities can produce atmospheres with above average levels of sulphur dioxide, chlorine, nitrous oxide and more toxic effluents and contaminants. A knowledge of the corrosion of metals, paints and other coatings is desirable e.g. anodic and cathodic inhibitors and adsorption and vapour-phase inhibitors. Outgassing from insulation and resin-bonded materials must also be considered.

The absorber and the cover are subject to exposure to dust which can lower the optical properties of the surfaces (Garg and Gupta, 1978)

Vacuum Stability

Many solar collectors, e.g. the evacuated tube collector are designed to reduce convective losses by mounting the receiver surface in an evacuated or partially evacuated low thermal conductivity inert gas enclosure. Such surfaces as the silicon-silver bilayer (Seraphin, 1976) and the $A\ell_2O_3$ - Mo - $A\ell_2O_3$ (AMA) selective interference filter (Schmidt, Park and Jannsen, 1973) have been tested to determine the maximum temperature at which they remain stable in vacuo. Surfaces that may oxidise readily in air at normal pressures may be capable of operation at higher temperatures in a controlled environment.

Compatability of the Coating/Substrate Interface

Together the coating and the substrate will have to withstand the combined effects of differential thermal expansion and electropotential differences. Any loss of adhesion can lead to a decrease in the efficiency of thermal conductance. Stress cracking opens the way for subsequent attack of the substrate and due consideration should be given to dimensional tolerances at plate fixing points. Figure 4 shows cracks in the surface of a copper oxide on copper selective tandem exposing the substrate to further oxidation and an eventual decrease in surface optical properties. The use of dissimilar metals such as a zinc absorber and copper substrate can allow the possibility of localised galvanic corrosion and loss of surface integrity.

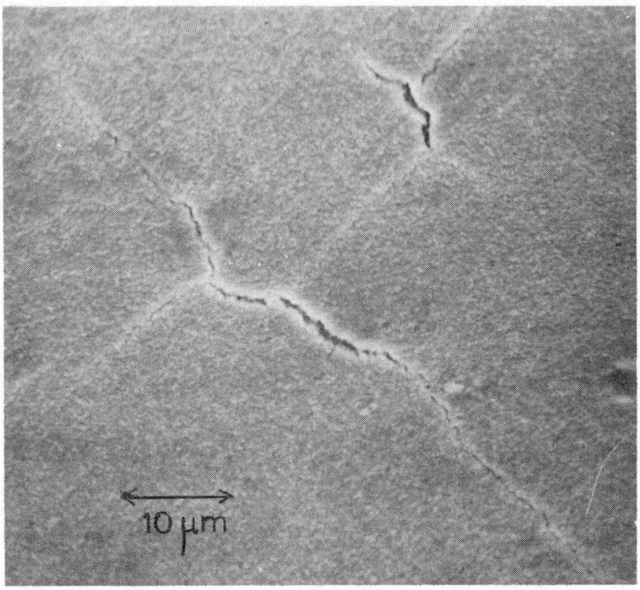

Fig. 4. Electromicrograph of Copper Oxide on Copper Selective Tandem showing Cracks in the Surface Film (x 2,000).

In the production of selective paints (Williams, Lappin and Duffie, 1963, Lin and Zimmer, 1977) where a minimum of binder is used in the coating composition a close examination of the binder/substrate relationship must be made.

After deposition of the absorber surface on the substrate the absorber plate is often shaped or formed to give the required design and the coating/substrate combination must be capable of withstanding the mechanical bending and deformation forces.

Other Considerations

Surfaces should be tested for abrasion and impact resistance and the effect of handling the surface should be studied.

In the case of the trickle collector the importance of the interaction between the working fluid and the absorber surface should not be neglected (Ashti A., 1979). Consideration must also be given to the repairability of the absorber coating since it may need to be maintained during its working lifetime. It is here that the selective paint offers an attractive advantage. Electrodeposited selective surfaces would need to be returned to a company for re-plating and non-selective paints, although readily re-applied result in significant radiant losses from the collector assembly.

The degradation considerations outlined in 3.1 - 3.6 are summarised in Table 1.

TABLE 1. Degradation Mechanisms and Issues of Concern Pertinent to Solar Absorber Coatings

High Solar Absorptance, α_s
Low Thermal Emittance, ε_h
High Thermal Conductivity
Maximum Operating Temperature
Temperature Cycling
Ultra-violet Irradiation
Humidity
Atmospheric gases and Contaminants
Vacuum Stability
Compatability of Surface Coating with Substrate
Shaping or Forming after coating
Repairability
Cost/Unit area
Impact and Abrasion Resistance
Effect of Dust on Optical Properties

DURABILITY TESTING

The diversity of degradation mechanisms of concern present a wide challenge to the materials scientist attempting to establish correlations of the relative lifetimes of surface coatings. It is inevitable that the coating will be subjected to a combination of different effects and conditions and any one or more of the previously mentioned mechanisms may be active in any degradation observed. In order to attempt to isolate causes from the macroscopic observable effects a series of tests must be carried out from which some quantifiable results may be extracted.

Natural Exposure Testing

One should ask the questions 'What nature of test will give a meaningful correlation with real life operation?' and/or 'Does the test induce the same effects both qualitatively and quantitatively as would occur in a real life situation from exposure to the natural environment of sunlight and weather?' Clearly the only test that can be absolutely relied upon to reveal the actual degradation and surface lifetime is that of natural exposure in a working solar system. Such a test is in progress at the Solar Energy Unit. The black chrome, copper oxide and non-selective matt-black paint surfaces which serve as the absorber coatings for the domestic water heating system installed in the students' residence at Llys-Tal-y-Bont (Taylor, 1979) are being monitored for variations in the optical properties of solar absorptance and thermal emittance. This programme has been underway for two years and preliminary results which reveal no observable changes in the black chrome and matt black surfaces and some increase in the thermal emittance of the copper oxide surface caused by further oxidation of the copper substrate and a consequent thickening of the surface film have been published (Hutchins, 1979a).

A summary of the results is presented in Table 2.

TABLE 2. Optical Properties of Absorber Surfaces as a function of In-service Lifetime of the Llys-Tal-y-Bont Solar Water Heating System.

Surface	Initial	Properties	12 Months		18 Months	
	α_s	ε_h	α_s	ε_h	α_s	ε_h
Black Chrome	0.96	0.16	0.96	0.16	0.95	0.16
Copper Oxide	0.87	0.10	0.88	0.15	0.89	0.12
Matt-Black Paint	0.95	0.90	0.95	0.90	0.95	0.90

Although such tests produce meaningful data from which informed evaluations may be made the rate of arrival of data is exceedingly slow and it would be of great assistance if accelerated weathering and ageing tests could be carried out. In addition the lack of a large number of sample surfaces render the results of Table 2 statistically difficult to interpret and highlight the need for large-scale programmes of testing. In-service natural exposure tests are extremely important since they provide the reference calibration to which all accelerated tests must eventually concur but the real problem to-day is to devise tests which simulate as closely as possible the in-service conditions and allow meaningful and helpful correlations and conclusions to be made.

Accelerated tests have been proposed (Riesch 1978, Gillett 1979) and may be classified into two distinct categories (1) qualification testing, and (2) ageing testing.

Qualification Testing

Qualification tests are designed to observe whether the candidate material will withstand the extreme situations which can occur in real-life operation. Examples are thermal shock caused by filling a cold collector with boiling fluid or abrupt freezing of the collector. Impact testing and abrasion testing, mechanical stressing and bending,and salt spray exposure can also be considered. This type of testing is also termed 'reliability testing' and allows identification at an early stage of materials that can have a high probability of basically attaining a long lifetime as an absorber surface coating.

Ageing Testing

Ageing tests are more complex than qualification tests. In an ageing test a specific degradation process, e.g. uv-irradiation, may be chosen for study and a test designed specifically to study a single cause. The difficulty arises when one wishes to draw a quantitative relationship between the accelerated or concentrated specific test and the in-service performance of the material or coating which is subjected to a combination of different causes. Nevertheless ageing tests provide an opportunity to perform laboratory experiments that will yield information in a time short compared to the in-service data acquired from natural exposure. Humidity or environmental chambers can be used to simulate conditions of elevated temperature or thermal cycling in the presence of moisture or atmospheric contaminants, such as SO_2, NO_2, and salt under a wide range of conditions. Test chambers can be used for the study of UV-irradiation in quantities greater than would normally be experienced under solar irradiation and facilities can be incorporated for the introduction of moisture etc.

Stagnation testing in which a solar collector is exposed to the climatic environment under conditions of 'no-flow', i.e. the collector carries no heat transfer fluid, are already in progress (Hayward, 1979). A similar idea has been applied to the exposure of solar absorber coatings under the present contract. A range of selective and non-selective coatings were chosen and a range of small samples (80x80mm) were prepared. The initial optical properties of these samples were measured and the samples were then mounted in a weather-tight, glazed box which was not vented in order to provide favourable conditions for the formation of condensation in the environment surrounding the surface coatings. The box with samples was then mounted on an outdoor test stand for a period of one year. A photograph of the samples undergoing this forced exposure testing is reproduced in Figure 5. At the end of the exposure period the samples were returned to the laboratory and their optical properties remeasured. A summary of the results of this test are presented in Table 3. A similar exposure test is underway at the National Bureau of Standards in the USA and the experimental arrangement is shown in Figure 6. Definitive correlations of relative surface durability and lifetime are not possible with so few results as are contained in Table 3 and further, enlarged tests are now in progress.

Fig. 5. Solar Absorber Coatings Undergoing Forced Exposure with No-flow Conditions (Cardiff, UK).

Fig. 6. Solar Absorber Coatings Undergoing Forced Exposure with No-flow Conditions (National Bureau of Standards, Maryland, USA).

In addition to the forced exposure test a small laboratory apparatus has been constructed to study the effects of accelerated thermal cycling. A schematic representation of this apparatus is shown in Figure 7.

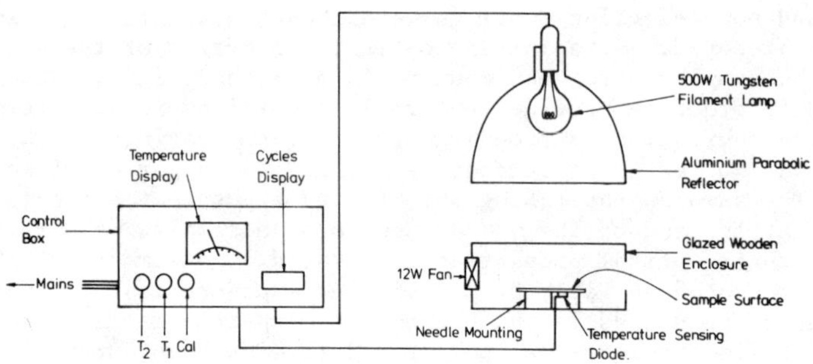

Fig. 7. Schematic Representation of Accelerated Thermal Cycling Apparatus.

The absorber surface to be tested is mounted in a glazed enclosure on three needle points which minimise conduction losses from the sample to the surrounds. The glazed enclosure can be cooled by a small 12W fan. The temperature of the surface is measured by a calibrated diode which is secured to the surface through a screw contact. The sample is heated by a 500W tungsten lamp housed in a parabolic reflector. The apparatus is controlled by an electronic circuit which allows the sample to be cycled between any two temperatures within the range 0-150°C.

Once the upper (T_1) and lower (T_2) temperature points have been set the apparatus may be switched on. If the sample temperature T_s is less than T_1 the lamp will be on and the sample will be heated. When T_s=T_1 the lamp is automatically switched off and the cooling fan is activated. The sample temperature falls until T_s=T_2 at which point the lamp once again switches on and the cycle is repeated. The number of cycles is recorded and displayed on the front panel of the control box. No quantitative results have yet been produced from this apparatus.

MEASUREMENT TECHNIQUES

For absorber coating durability performance evaluation monitoring of the optical properties before and after any exposure test is used as the reference guideline. However the discovery of a change in the solar absorptance or thermal emittance of the coating cannot directly reveal the mechanism responsible for any observed variations. Here the scanning electron microscope is invaluable for the study of structure and surface changes. SEM can reveal cracks, blisters, loss of adhesion, changes in film thickness, or density of voids together with microstructural changes that may be caused by oxidation or UV- irradiation, e.g. an increase in average

TABLE 3 Optical Properties of Solar Absorber Coatings before and after one year of forced exposure under no-flow conditions.

Sample	Unexposed Optical Properties		Optical Properties following one year no-flow exposure	
	α	ε	α	ε
Black Chrome on Aluminium	0.95	0.39	0.95	0.46
	0.95	0.40	0.96	0.38
	0.95	0.36	0.96	0.46
	0.94	0.38	0.94	0.44
	0.96	0.41	0.95	0.45
	0.94	0.36	0.95	0.35
	0.95	0.36	0.95	0.51
	0.95	0.37	0.96	0.40
	0.96	0.38	0.96	0.38
	0.96	0.39	0.94	0.45
	0.96	0.42	0.96	0.47
Black Chrome on Copper	0.94	0.48	0.95	0.71
	0.94	0.51	0.95	0.56
	0.94	0.48	0.94	0.82
	0.94	0.44	0.93	0.48
Black Chrome on Copper	0.92	0.16	0.93	0.20
	0.93	0.15	0.93	0.17
	0.93	0.16	0.94	0.20
	0.95	0.16	0.92	0.16
	0.94	0.18	0.94	0.21
Black Chrome on Copper	0.92	0.11	0.93	0.11
	0.94	0.12	0.95	0.13
	0.94	0.10	0.92	0.11
	0.94	0.12	0.93	0.12
	0.93	0.12	0.92	0.13
	0.94	0.12	0.93	0.15
	0.93	0.14	0.93	0.17
	0.93	0.14	0.94	0.13
	0.94	0.14	0.94	0.14
Anodised Aluminium	0.95	0.90	0.95	0.90
	0.95	0.88	0.95	0.90
	0.95	0.88	0.95	0.85
	0.95	0.85	0.95	0.86
	0.95	0.90	0.94	0.88
	0.95	0.88	0.95	0.89
	0.95	0.85	0.94	0.85
	0.95	0.85	0.95	0.87
	0.95	0.85	0.95	0.87

TABLE 3 continued:

Sample	Unexposed Optical Properties		Optical Properties following one year no-flow exposure	
	α	ε	α	ε
Chemically Oxidised Stainless Steel	0.88	0.14	0.89	0.14
	0.88	0.16	0.88	0.18
	0.76	0.16	0.77	0.14
	0.75	0.15	0.77	0.16
	0.69	0.15	0.67	0.15
	0.88	0.16	0.88	0.15

All numerical values ± 0.03

particle grain size, lattice imperfections, or porosity of the surface layer. Changes in chemical composition can be studied using electron spin chemical analysis (ESCA), X-ray photoelectron spectroscopy (XPS), infra-red spectroscopy, or Auger depth profiling. One investigative tool that is always present and often overlooked is visual inspection and care must be taken to ensure that this simple evaluation is always made before embarking on any of the above mentioned more sophisticated inspection tools.

DISCUSSION

The quantitative relationship between accelerated testing and in-service lifetime is not easy to correlate. Degradation mechanisms are widely different in character and nature and investigative tests must be based upon the important properties and exposures expected in service. A full-scale durability and materials evaluation programme is highly demanding. The limited number of samples tested to date make it unlikely that any definitive correlations can be accurately stated. In order to be able to accommodate statistical variations in any one surface coating large numbers of surfaces must be prepared possibly using different production parameters in order to study anciliary effects such as electroplating bath compostion and the subsequent effect on optical properties and thermal stability. Since the number of samples is likely to be large and the measurements of a repetitive nature data-logging equipment for the determination of temperature changes in the samples and the time spent in any given temperature band, and portable instrumentation for the measurement of surface optical properties are both desirable, together with the capacity to produce a large volume of surfaces under a range of different conditions. Non-destructive tests are also a priority since, following testing and examination, it may be desirable to return the sample for further exposure.

Looking to the future it is desirable that efforts be made to establish a materials data base that will serve to provide information on durability, maximum operating temperature, materials resource limitations, industrial availability and cost/unit area in addition to fundamental information on the nature of a material and its optical properties.

At the time when definitive correlations can be made between accelerated testing for degradation processes, such as uv-irradiation and thermal cycling in the presence of moisture and contaminants, and in-service performance data, it will then be appropriate to define standard means of testing and evaluation. Until that time arrives efforts should be directed towards producing a quantity of materials durability data that provides some information on the behaviour of surfaces and materials in in-service conditions. Durable, cheap and maintainable materials mean cost-effective, commercially acceptable solar energy systems. The materials scientist should be prepared to devote more time and attention to the identification of such materials than has previously been given in the past.

REFERENCES

Ashti, A., M.Sc. Thesis, 'Performance Analysis of a Trickle Solar Collector' University College, Cardiff (1979).

Call, P., 'National Programme Plan for Absorber Surfaces R & D', SERI/TR-31-103 (1979).

Garg, H.P., and Gupta C.L., 'Flat-Plate Collector Studies and Design Data for India', Proc. ISES Congress, New Delhi, (1978)

Gillett, W.B., 'Reliability and Durability of Solar Heating Systems', SEU Report No. 154, (1979).

Hayward, K., 'Collector Testing in the Solar Energy Unit', Helios 5, pp 6-9, 1979.

Hutchins, M.G., Rawcliffe, R., Gillett, W.B., and Green, A.A., 'Flat-Plate Collectors and Solar Water Heating Systems', SEU Report No. 160, (1979).

Hutchins, M.G., 'Thin-film Coatings in Solar Water Heating Systems', Proc. Future Energy Concepts pp 249-252, IEE, London (1979a).

Hutchins, M.G., 'Current Advances in the Development of Electrodeposited Black Chrome Surfaces', Helios 7, pp 6-9, (1979b)

Hutchins, M.G., 'Optical Measurement Techniques for the Characterisation of Solar Absorber Coatings', SEU Report No. 159, (1979c).

Ignatiev, A., O'Neill, P., and Zajac, G., Solar Energy Materials, 1, 69, (1979)

Lin, R.J.H. and Zimmer, P.B., Honeywell Report No. COO-2930-12, Minneapolis, (1977).

Peterson, R.E., and Ramsey, J.W., J. Vac. Sci. Tech., 12, 174, (1975).

Pettitt, R.B., and Sowell, R.R., 'Thermal Ageing Characteristics of Electrodeposited Black Chrome Solar Coatings', ISES Silver Jubilee Congress, Atlanta, Ga.,(1979).

Riesch, G., Qualification Testing and Ageing of Solar Equipment, CEC/Ispra, (1978).

Schmidt, R.N., Park, K.C. and Janssen, J.E., 'High Temperature Solar Absorber Coating Study', AFML-TR-73-80, (1973).

Seraphin, B.O. and Meinel, A.B., Optical Properties of Solids-New Developments, North Holland Publishing Company, Amsterdam, (1976).

Seraphin, B.O., Thin Solid Films, 39,78, (1976)

Sikkens, M., and Francken, J.C., 'Spectral Selective Properties of Nickel Carbide with Varying Carbon Content', Proc. ISES Silver Jubilee Congress, Atlanta, Ga., (1979).

Tabor, H., Bull. Res. Counc. of Israel, 5A, 119 (1956).

Taylor. P., 'Monitoring Solar Water Heating Systems' Helios 6, pp 1-5, (1979).

Williams, D.A., Lappin, T.A., and Duffie, J.A., J. Eng. Power, 5, 213, (1963).

ACKNOWLEDGMENT

This work was funded on a grant from the Science Research Council.

STATE POLICE BARRACKS AT BAGNOLS-SUR-CEZE WITH SOLAR HOT WATER AND SOLAR SPACE HEATING SYSTEMS

G. Chouleur

Solar Heating Engineer, 6 rue Fresque, 30000 Nimes, France

ABSTRACT

An account is given of the various design and constructional aspects of the solar heated State Police Barracks at Bagnols-Sur-Ceze. Information is also provided on local meteorological conditions, the building's heating requirements and solar contribution. Aesthetic and installation details concerning the solar collectors integration into the building fabric are also discussed.

KEYWORDS

Solar heated barracks; buildings; temperature; solar radiation; thermal insulation; Chouleur patent; collectors; storage; controls.

1. INTRODUCTION

The solar-heated state police barracks at Bagnols-sur-Ceze is the sequel to the construction of five houses making partial use of solar heat at the EDF (French Electricity Authority) housing complex at Aramon and three houses, designed for complete solar heating, on the site of the conventional power station built there.

These houses were extensions of the techniques used hesitantly for solar water heating in the villa of Mr. Chatelain, and the subsequent and more sophisticated solar equipment used for partial heating and hot water in the villa of Mr. Cherence.

The buildings ar Aramon, in particular the houses on the power-station site, have been visited by engineers and economists from all parts of France and from throughout the world, all of whom were astonished by the scope of this experiment and curious as to its cost and results.

While the houses of the residential complex drew some criticism as regards the colour and locations of the collectors, those built on the power-station site calmed the architect's fears. The whole world is, or should be, aware of the energy shortage threatening us. The whole world knows that none of the mineral forms of energy (coal, oil, even uranium) are inexhaustible.

On the other hand, renewable energy sources such as the sun, tides, and the wind are inexhaustible. Their value is even greater first because they can postpone the onset of the shortage and then because they can release petroleum fuel for other uses for which it is currently irreplaceable (automobiles, aircraft, petrochemiclas, etc...). Finally the course that research is now taking suggests a substitution of conventional energy sources by low impact ones, thus putting a stop to the cruel and dangerous struggle for possession or control of oil-rich zones and the more insidious but no less dangerous struggle of the markets. These vital appetites and confrontations could bring war within 20 years.

The solar-heated state police barracks at Bagnols-sur-Ceze were designed along the same lines as the earlier projects, but benefitting from the errors made and from the new knowledge acquired in the meantime.

One of the repreaches levelled at the five houses of the Aramon residential complex was the coldness of the black mass of the collectors. Alas, these collectors had been added to a house that had already been designed. This reproach was not made in the case of the three houses on the power-station site, because of the location of the collectors and their very modest decoration on the glass panels. Another concern was the duration of heat storage and the difficulties of using it.

The solar-heated state police barracks at Bagnols answers these misgivings:

- The collecting system is made decorative throughout by means of stepped collectors, using the Chouleur patent (patent No 76.39215 and No US patent 79.17240), and the high-aspect-ratio, controlled-stratification tank built under another Chouleur patent (patent No 77.23572).
- The wall on which the collectors are mounted becomes an architectural element.
- The large temperature difference between the top and bottom of the tank makes it possible, through the use of probes and electronically controlled valves, to tap the heating fluid at the point closest to the desired temperature, and so economise the heat energy stored.

The client for this project was the Department of Gard (Groupement de Gendarmerie du Gard, Gendarmerie de Bagnols-sur-Ceze). The architectural and solar heating study was carried out by the author and his office at Nimes. A subsidy of 135,000 FF was granted to the Department of Gard by the Agency for Energy Savings for implementing the solar part of the project.

2. BRIEF DESCRIPTION OF THE BUILDING COMPLEX

The Site:

The administrative and residential buildings of the state police barracks are built on a 9,290 m^2 site given to the Department of Gard by the City of Bagnols-sur-Ceze.

The buildings are located in front of the departmental road No 6 from Ales to Bagnols, a road which has been widened to 12 meters under the city development plan, and on the Rue des Aubepines, which serves some recently built housing.

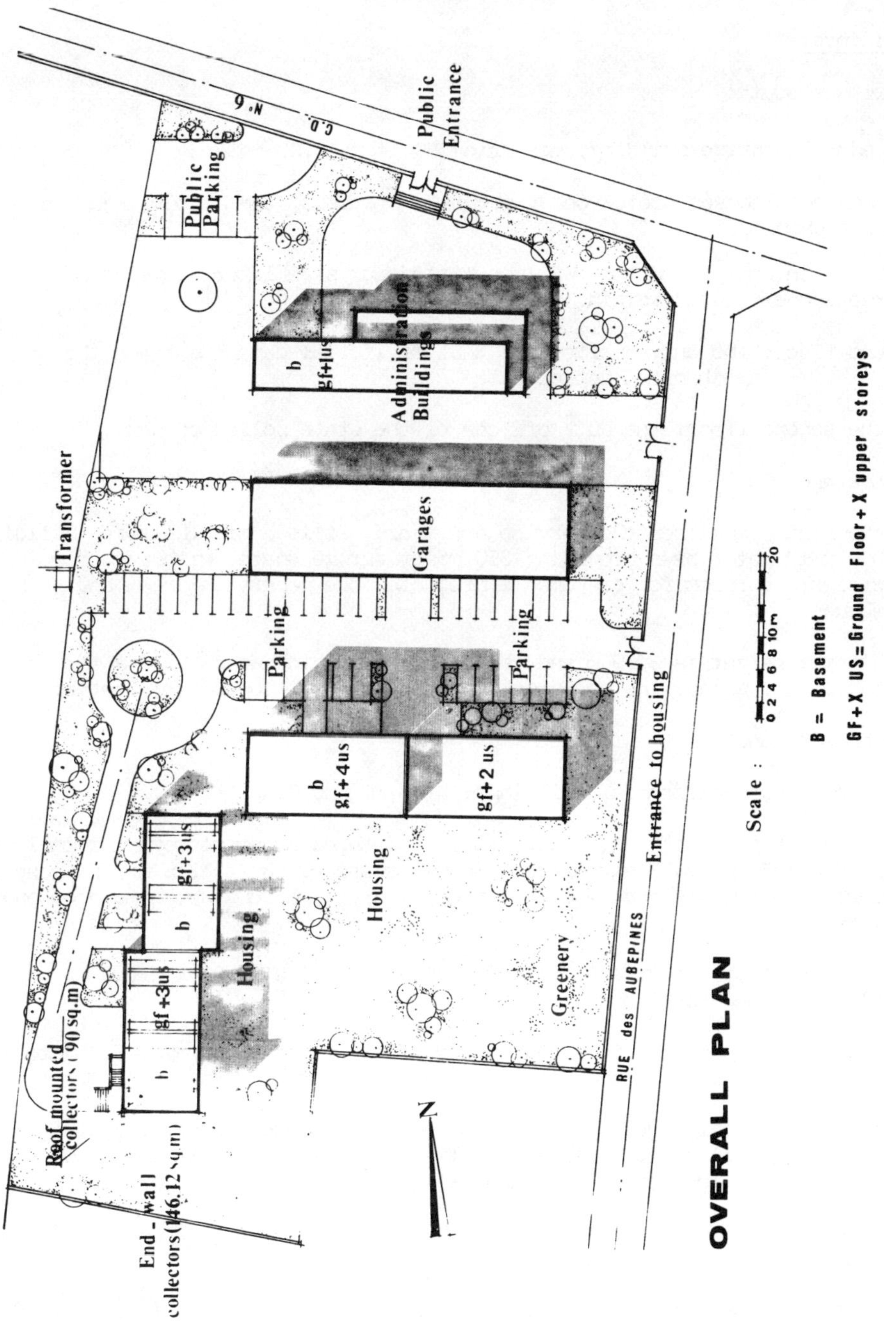

Fig. 1 Overall Plan

Its longest dimension is from north to south; it slopes gently from north to south and slightly overhangs the departmental road.

The Buildings:

From north to south:

1) The administrative building, measuring 36.14 x 10.29 meters

   The basement houses store rooms, security premises, the battery room and the switchboard.

   There is an empty room that will be fitted out at a later stage as a recreation or light training room.

   Ground floor: the duty quarters of the territorial detachment and the mechanized detachment.

   On the second floor: the duty offices of the state police company.

2) The garage

   This building encloses the service courtyard. It is a single-storey building with 6 x 12 meter bays providing 350 m^2 of garage space, with fuel pump-meters, store rooms for consumable products, the generating room and a warehouse.

   The lintel height is 3.50 m and the clear height inside is 3.90 m at the bottom of the slope. The building is closed by roll-down metal shutters.

3) The housing units

   There are four residential buildings: A, B, C and D.

   Building A has a basement, ground floor and three upper storeys and includes eight housing units. Its outside dimensions are 20.53 x 11.15 m. The basement includes the boiler room, a common room, a dryer, a rubbish chute room and cellars.

   Building B is of the same height. Its dimensions are 17.56 x 11.26 m. It also includes eight housing units.

   Building C has a basement, ground floor and four upper storeys. Its dimensions are 10.71 x 11.26 m. It includes ten housing units.

   Building D has only two upper storeys above the ground floor. Its dimensions are 21.31 x 11.16 m. It includes auxiliary housing units with dormitories, a guard room, a recreation room and cellars.

The walls of these buildings, except for the garage, are made of Menerbe freestone (Luberon Massif).

Special attention has been paid to the thermal insulation, which includes, after the freestone, a 2 cm air gap, a 6 cm panel of polyurethane foam and 10 mm plasterboard.

The intersections of the ceilings and floors with the outside walls have also been suitably insulated. There are composite floors throughout and the buildings have flat roofs. The bottom floor and flat roof are both insulated by polyurethane foam. The conventional weatherproofing is protected by a 6 cm thick layer of gravel.

The windows are made of wood, with slatted shutters. Each housing unit has a large loggia on the main facade, in front of the living room, and a similar loggia in front of the kitchen.

Buildings A and B are oriented north to south and so face east and west.

The south wall of building A is the wall on which the collectors are mounted. There are also collectors on the flat roofs of buildings A and B. Building A is offset with respect to building B.

Buildings C and D are oriented east to west and face north and south.

Together they form an L open to the south-east, bounding an area with a playground and gardens. In front of the administration building is the building's entrance courtyard, with "users" parking. The service courtyard is between the administration building and the garage. A covered passage through the garage building allows the policemen to go directly from the duty areas to the residential areas. Another space between the garage and buildings C and D contains parking spaces for the policemen. Gardens surround and brighten up the area. Altogether, the residential buildings provide:

- 2 officers' housing units, 28 policemen's housing units, and 2 four bed rooms for auxiliary policemen;

- 1 NCO's room, 1 off-duty room and sanitary facilities.

The total built-up area (outside measurements), including the garage, is 5,133 m^2. The usable office space is 446,58 m^2. The total living area of the housing units is 2,363 m^2.

To provide an attractive appearance while remaining within the budget allocated, simple lines were chosen for all of the buildings, made of thick freestone, with monotony avoided by the differences in alignment and height of the buildings, the setback and colour of the loggias and the projection of the balconies.

Table 1 General Climatic Data

1 - LOCATION:

The solar heated state police barracks is located in the municipality of Bagnols-sur-Ceze (Gard). This town is classed in climate zone C. Its latitude is 44°10' North.

2 - WIND VELOCITIES: region III, standard NV.65.

- Normal values 38.3 m/sec or 137.9 km/h
- Extreme values 50.7 m/sec or 182.5 km/h

3 - OUTSIDE TEMPERATURES:

- Annual maximum + 28°C
- Annual minimum - 4°C
- Annual mean + 15°C

4 - WATER:

TH of water: 20 to 35° (hard water), or 200 to 300 grams of calcareous matter per cubic meter of water.

5 - SOLAR RADIATION

There is an average of 2,673 hours of sunshine per year, with a total incident solar radiation (on 63° inclination) of 1775.3 kWh/m^2/year.

Table 2 General Design Data

Design Requirements

For an outside temperature of $-5^{o}C$, the comfort temperatures required in the building are:

- Bathroom	+ 22°C
- Living room, kitchen	+ 20°C
- Bedrooms, inside passages	+ 18°C
- Entrance	+ 18°C
- Public, radio and working rooms and offices	+ 20°C
- Corridors and sanitary facilities	+ 18°C
- Store rooms and archives	+ 16°C

Hot Water Requirements

Estimated requirement of hot water at design temperature of 55°C (distribution temperature of 40°C) 250 l per housing unit per day, resulting in an energy requirement for the 31 housing units of:

$$\frac{250 \times 31 \times (55-10)}{1,000} = 348.75 \text{ calories/day or } \underline{\underline{405.6 \text{ kWh per day}}}$$

Thermal Insulation

	(heat loss per unit area of wall)
- Outside walls	$K = 0.46\ w/m^2\ {}^oC$
- Flat roof	$K = 0.43\ w/m^2\ {}^oC$
- Floor	$K = 0.51\ w/m^2\ {}^oC$
- Double glazing	$K = 3.2\ w/m^2\ {}^oC$
- Single glazing	$K = 5.7\ w/m^2\ {}^oC$

The water supply temperature was estimated at 10°C, and the hot water feed temperature into the distribution network 52°C. These estimates include a large safety margin.

Double glazing was used on the north and east facades of the housing units and offices.

The offices and apartments are equipped with controlled mechanical ventilation.

Table 3 Heating Requirements and Solar Contribution

Site: Bagnols-sur-Ceze - Collectors facing south, 63° from horizontal. Collection area 272 m^2.

			REQUIREMENTS					
No	Month	Days	Space Heating (kWh)	Hot Water (kWh)	Total (kWh)	Solar Contribution (kWh)	Balance (kWh)	Solar Coverage (%)
01	January	31	83 472	12 574	96 046	14 658	81 388	15
02	February	28	68 357	11 357	79 795	18 866	60 929	24
03	March	31	52 282	12 574	64 857	33 871	30 896	52
04	April	30	32 312	12 168	44 480	43 674	806	98
05	May	31	11 444	12 574	24 018	48 205	–	100 (1)
06	June	30	–	12 168	12 168	51 904	–	100 (1)
07	July	31	–	12 574	12 574	58 609	–	100 (1)
08	August	31	–	12 574	12 574	47 604	–	100 (1)
09	September	30	–	12 168	12 168	33 663	–	100 (1)
10	October	31	20 868	12 574	33 442	27 605	5 837	83
11	November	30	54 975	12 168	67 143	18 010	49 133	27
12	December	31	76 068	12 574	88 642	13 642	74 863	16
			399 860	148 047	547 907		303 942	

When allowance is made for the occupancy factor with 12.5 % free heat provided by the occupants, the actual balance to be made up is 87.5 % of the value found, or $\frac{303,942 \times 87.5}{100}$ = 265,949 kWh/year.

This amounts to a saving of: $\frac{281,958}{547,907} \times 100 = 51.4$ %

(1) Except during extended periods of bad weather.

3. THE INSTALLATION

3.1. Solar Energy Collection

Solar energy is collected by flat collectors sloped at 63°C from the horizontal made of extruded aluminium forming ribs enclosing copper tubes that are expanded after insertion.

The outside dimensions of the elements are 2.054 x 0.360 m. The shell is made of galvanised sheet.

The back and sides are protected by a 2.5 cm layer of polyurethane. The exposed portion is protected by panes of ordinary 4 mm glass held by frames. In view of the slope and small size of the panes, damage by hail is relatively improbable.

The collectors on the bottom of the south facade (2 x 13 m) have outside dimensions of 1 x 2 m and are placed vertically. In these collectors, a small strip of aluminium divides the area of the pane into two equal 1 m^2 parts. These collectors are made of steel. They are commercially available radiators, and are likewise placed in galvanised sheetmetal frames. The pane is again secured by frames, with an insulating compound. Because they are mounted vertically, damage by hail seems relatively unlikely.

The collector array includes:

(a) On the south end of building A, stepped vertical panels with overall dimensions of 14.70 x 10.60 m, consisting of 27 rows each containing 7 elements with a working area of 0.66 m^2 each, arranged as per the Chouleur patent (U.S. patent No 75.39215) and, at the bottom, 13.2 m^2 collectors of the second type described above, placed upright (see Figs 2 + 3)

The collectors placed more than 2 m above the ground, in a stepped array, have decorated undersides. These undersides are detachable and can be removed for reconditioning of the paint or for redecoration. They represent a gold sun against a red and blue background.

The flat vertical collectors at the bottom retrace the history of the French state police uniform from 1715 to the present.

As it can be seen, the collectors, rather than being a constraint to the architect, they become here a decorative element. Art is restored to its place in building and the painter can finally express himself on the very outside walls of buildings.

Because the undersides can be removed all artwork can be prepared in the studio.

(b) On the flat roof, a set of five rows of collectors, 2 m high and 9 m long

These collectors are mounted on stands that are themselves fastened to a concrete slab. These stands are heavily wind-braced to withstand the force of the wind.

In front of these collectors, there is a reflecting surface consisting of a plane mirror isolated from the concrete slab to which it is fastened by a 5 cm air gap. This reflecting surface increases the efficiency of the collectors by about 40 %.

Fig. 2 The South Face of the Bagnols-Sur-Ceze Building Covered With Collectors

Fig. 3 A Close View of the South Wall, Vertically Mounted, Collectors.

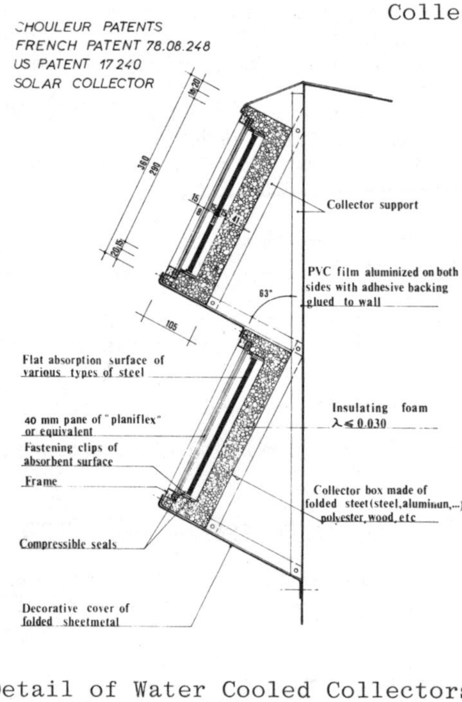

Fig. 4 Mounting Detail of Water Cooled Collectors on the South Face of Police Barracks Building at Bagnols-Sur-Ceze (All Measurements in Metric)

3.2. Storage

The decorative collectors are not the only original feature of the solar heating installation of the Bagnols state police barracks.

The solar storage tank, which is very slender (i.e. has the highest possible ratio of height to diameter), has inlets and outlets at various heights and convection brakes consisting of truncated cones topped by a cylinder guilding the hot water from the center of the tank to the top and allowing the cooled heating fluid to descend along the walls of the tank.

The solar hot water coming to the tank via the heat exchanger is stored in this 20,000-liter tank, which consists of a reinforced concrete pipe with a steel core. The steel core is covered inside and out by a layer of suitably reinforced cement.

The steel barrier renders the tank perfectly watertight. The concrete is cast on the steel in a single operation to produce a uniform tube in which the reinforcements are perfectly coated, without discontinuity. The prestress is constant over the entire height of the tank, both lengthwise and radially.

The bottom of the tank is also made of concrete and the top is made of steel. The ends have a basket-handle profile. The inside diameter of the tank is 1.90 m and its inside height in the centerline is about 7 meters. The ratio of height to diameter is 3.68. The outside of the tank is insulated by 12 cm of polyurethane foam. The bottom of the tank is set on 16 cm of high-density polyurethane foam, and the top is also insulated. The weight of the empty tank, fully equipped, is 24 tons.

The water from the two arrays of collectors is recovered by a single header which forms part of the primary circuit. Then through a heat exchanger water is fed to the storage tank via two points (upper and lower) where three way automatic valves are fitted.

The water of the heating circuit is also drawn from two points on the tank. An electronic differential control unit is responsible for the point selection process.

The devices equipping the tank are covered by a Chouleur patent (U.S. patent No 77.23572).

Cold drinking water is fed in at the top of the tank and descends via a 40 mm dia. tube to the bottom of the heat exchanger. It then flows up through the entire stainless-steel exchanger, which consists of three coils formed by two coaxial rows, 1.60 and 1.40 m in diameter, respectively, of type 22 CN-18.10 40 mm dia stainless steel tubes. The turns are 115.6 mm apart in the bottom coil, 77 mm in the middle coil, and 57.8 mm in the top coil. The bottom coil contains 94.20 m of tubes, the middle coil 141.28 m, and the top coil 188.33 m, for a total of 423.81 meters of tubes and a capacity of 534 liters.

STORAGE TANK

1 Temperature probe at top of tank
2 Temperature probe at 2/3 height
3 Temperature probe at 1/3 height
4 Temperature probe, fluid going to collectors
5 Outside temperature probe
6 Temperature probe, fluid leaving collectors
7 Upper heating temperature probe
8 Lower heating temperature probe
9 Heating return temperature probe
10 Hot-water inlet temperature probe
11 Hot-water delivery temperature probe
12 Temperature probe in utilities room

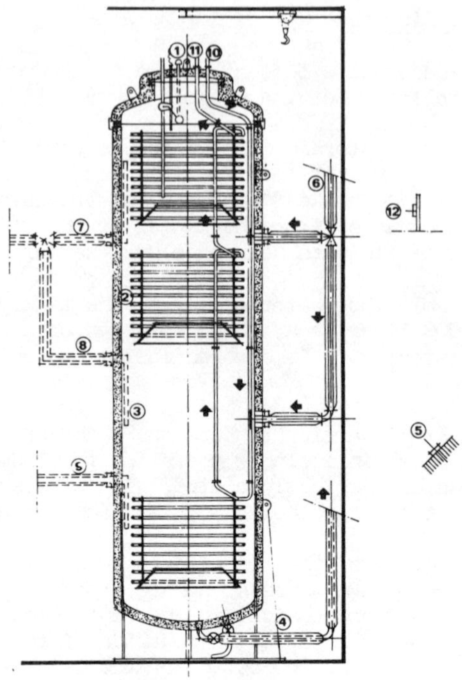

Fig. 5 Storage Tank

CIRCUIT DIAGRAM
E.R.E.N. G. Chouleur FRANCE

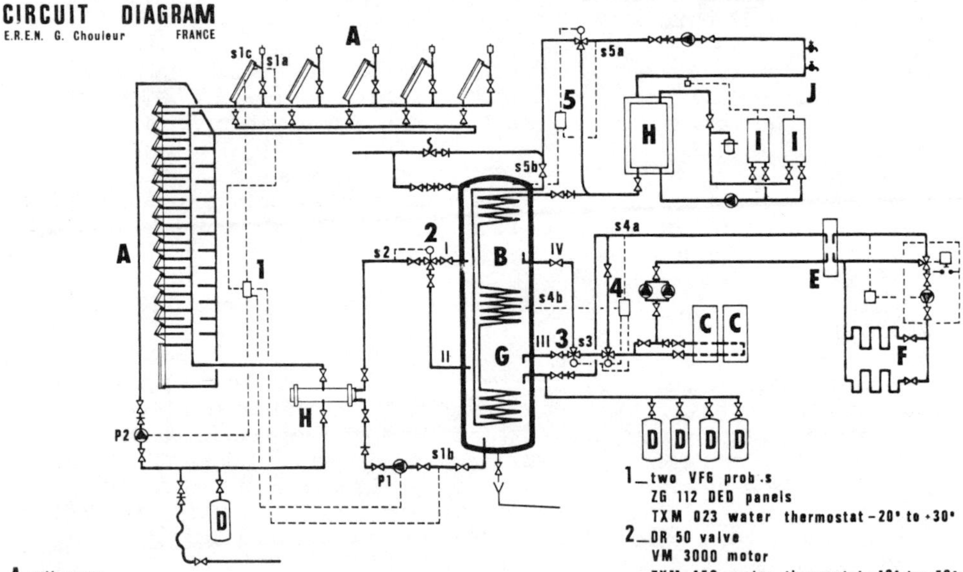

A collectors
B 20 000 liter solar storage tank
C gaz-fired auxiliary boilers with 2 water thermostat (heating)
D expansion vessels
E mixing cylinder
F heating floor
G hot-water coil
H exchangers
I auxiliary boilers
J use

1_two VF6 prob.s
ZG 112 DED panels
TXM 023 water thermostat -20° to +30°
2_DR 50 valve
VM 3000 motor
TXM 150 water thermostat +10° to +50°
3_TXM 490 water thermostat +40° to +90°
DR 50 valve
VM 3000 motor
4_two VF 6 probes
DR 50 valve + VM 3000 motor
ZG 112 DED panels
5_two VF 6 probes
ZG 112 DED panels
KMR 32 three-way valve
VMA motor

Fig. 6 Circuit Diagram

The collectors on the roof are preactically invisible from the ground level.

The collectors of both arrays are connected in parallel in a so-called "Tickleman loop".

There are automatic air vents at the high points of the rows of collectors on the roof.

In the case of the collectors on the south wall, the heating fluid is collected in steel tubes, carefully insulated and protected by galvanized sheetmetal boxes. The heating fluid from the collectors on the roof is also collected in steel tubes that are carefully insulated and protected from atmospheric corrosion.

There are two closed circuits, with water as the heat transfer fluid, coupled to heat exchangers. The domestic hot water space heating and collection loops are shown in Fig. 6. Temperatures in the collection circuits range from 25°C to 80°C.

The working areas of the collectos are:

- on the south wall:	- stepped 26 x 7 x 0.66 =	120.12
	- flat 13 x 2	26.00
- on the roof:	- five 18 m^2 rows = 90.00	90.00
	- area corrected for effect of reflectors: 90 x 1.4 =	126.00
Total working area of collectors =		272.12 m^2

NOTES ON 3.1.

(a) Metracrylate was not used to cover the collectors because it deteriorates when exposed to ultraviolet radiation at a rate that depends on its own absorption in this portion of the spectrum, its concentration of photosensitizing impurities, and the reactivity of its structural pattern.

(b) The plates of metacryclate polyane used at Aramon are sufficiently stable with respect to solar ultraviolet radiation, but their coefficient of expansion is too high. The breaks that have been found in these plates resulted not from impacts or thermal shocks but from their expansion and contraction.

(c) In view of the strength of the wind at Bagnols-sur-Ceze, it may be necessary to protect the backs of the collectors on the roof by covering the acute angle with an insulated partition.

3.3. Note on Control Functions (see Figure 6)

The control functions are numbered from 1 to 5, while the letter S is used to indicate the various probes.

Control No. 1 - Solar collectors

Switches pumps P1 and P2 on and off.
Differential control. Temperature difference: 8°C
The pump operates if the temperature of S1a is greater than the temperature of S1b + 8°C.

Anti-freezing safety:

S1c is set to - 10°C; in other words, if the temperature of S1c is less than -10°C, pump P2 automatically starts. The primary circuit fluid is made of water and 30 per cent of antifreeze. Then the freezing point is -17°C.

Control No. 2 - Storage tank outlet to p

Control by three-way valve operated by probe S2 is set to a temperature of +40°C. If the temperature at S2 is over 40°C, the solar heated fluid enters via the upper point at I. If the temperature at S2 is below 40°C, the fluid is taken at the upper

Control No. 3 - Storage tank outlet to primary heating

Control by three-way valve operated by probe S3 is set to +45°C.
If the temperature at S3 is over 45°C, the fluid is taken via the lower point, at III. If the temperature at S3 is below 45°C, the fluid is taken at the upper point, at IV.

Control No. 4 - Primary heating

Differential control is set to 8°C. If the temperature at S4b, the probe at the centre of the tank, is greater than S4a +8°C (return from heating), the three-valve connects the primary to the storage tank.

If the temperature at S4b is less than S4a + 8°C, the three-way valve establishes a closed circuit to the gas-fired boilers.

These two boilers, cascade-mounted, are Saunier Duval type 7100 105,000-Kcalorie modular boilers (CC on diagram); refer to the attached data sheet.

Control No. 5 - Hot water loop

Differential control is set to 8°C.

If the temperature at S5b is greater than S5a + 8°C, the hot water return is via the exchanger of the solar storage tank.

If the temperature at S5b is less than S5a + 8°C, the hot water return is via the instantaneous exchanger heated by the gas-fired boilers, cascade-controlled by their own water thermostat.

These boilers are variable-control 30,000-Kcalorie boilers similar to the type SD 230.

AN OVERVIEW OF WORK CARRIED OUT BY SOLAR ENERGY DEVELOPMENTS

D. Michaelis

Solar Energy Developments, Bay 10, 16 South Wharf Road, London W2 1PF, UK

Solar Energy Developments (SED) is a London based consultancy set up in 1974 by Dominic Michaelis and Steven Szokolay. Its purpose is to carry out research and development, to offer a general consultancy service and to design solar energy installations and components.

System development was carried out initially for the Calor Group Ltd in conjunction with Fulmer Research. SED designed and supervised the building of a Solar Laboratory near Slough; the objective was to provide a test vehicle for a solar heat collection storage and distribution system to be developed jointly by Calor, Fulmer and SED. The solar collectors were to be tested at varying angles of elevation in order to simulate varying standard roof pitches. The test vehicle was also to allow comparison of performance at varying orientations. A multi storey stack storage system with superimposed water tanks to include a back-up boiler, hot water tank and a warm air distribution system was to be provided.

The laboratory has a steel base structure on castors that allows it to be moved about a pivot point forming the base of a service distribution core. The superstructure is a lightweight steel frame with glazing to the roof where it forms the cover to the solar collectors and also to the low front wall. Varying degrees of insulation are achieved by varying the internal linings to the glass. Effects of varying amounts of south facing openings can be studied, and detached, semi-detached and terrace house situations simulated by varying side wall insulation. The blank back wall is of corrugated steel sheeting backed by suitable urethane insulating material. The laboratory was completed in 1976 and is carrying out tests on panels and systems with a view to developing a commercialised solar domestic heating package for space heating and domestic hot water in the UK.

A system derived from these tests has been installed in housing at Milton Keynes, for which SED acted as architects. This project, aptly called Solar Court, is a joint experimental development by John Laing Research & Development Ltd and the Calor Group with funding from the UK Department of Energy. The aim of the project is to assess the comparative performance and viability for speculative housing of two different solar active space and water heating systems comparing them with a conventional house type. Three different house types are presented, three of each type.

Fig. 1 Solar Laboratory in Slough, U.K.

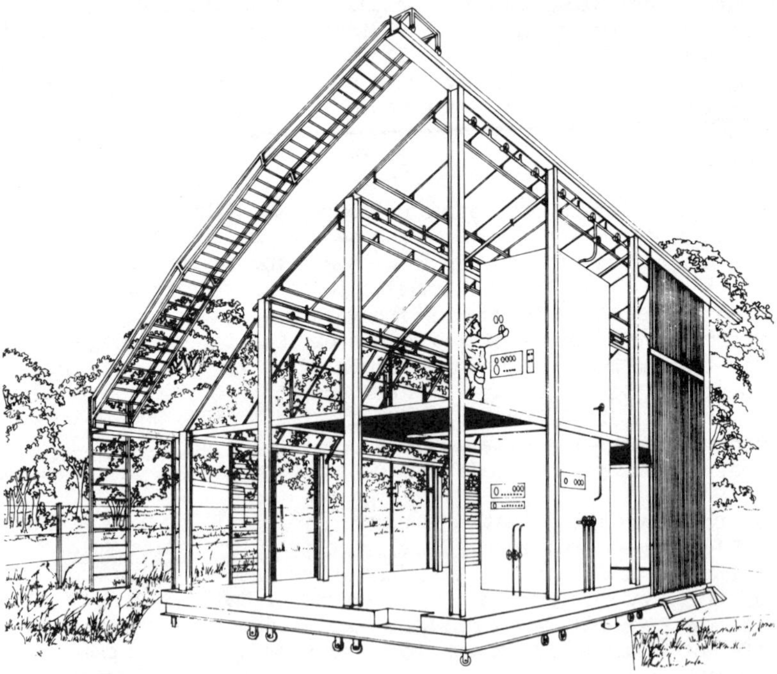

Fig. 2 Solar Laboratory for Calor at Slough

1. Monitor houses. These are well insulated double glazed detached houses of the same basic construction as the solar houses. They have large south facing ground floor openings to increase passive gains.

2. Solar Houses I: John Laingh R & D System. These have 18 m^2 of solar collectors on the lower roof pitch and are fitted with a solar assisted heat pump system. A large underground water store and waste water store are linked by the heat pump to a smaller day store. Background space heating is provided by floor coils with a supplementary warm air system.

3. Solar Houses II: Calor Group Ltd System. These houses have 40 m^2 of solar collectors on the upper and lower roof pitches. A water storage system is provided, space heating being by warm air. Air ro air and water to water heat recovery systems are included.

SED have also been working on a brief from Taylor Woodrow Ltd as consultants on a passive solar architectural project. The aim was to demonstrate that passive solar homes can be constructed in the UK to replace conventional estates at minimal extra cost; to quantify the reduction in energy usage due to the passive solar gains; and to provide design recommendations for future passive solar architecture in the UK.

Five houses on a typical domestic housing site, each of the four bedroom detached type based on a standard Taylor Woodrow design, were assessed for the project.

1. Direct Gain House. A large south facing glazing area was proposed with night insulation and horizontal shading elements to control summer gains. The building mass was an important variable which was adjusted for optimum performance.

2. Conservatory house. It was proposed that this house incorporates a conservatory one and a half storeys high on the south facing wall. The conservatory air space temperature is not critical to the comfort of the house and so gives control of solar gains. Thermo-circulation of conservatory air and night insulation provide additional control.

3. Trombe Wall House. The south facing wall is altered to provide a trombe wall. Night insulation and possibly selective surfaces are suggested to improve the potential for useful solar gain.

4. Water Wall House. It was proposed that the south facing wall be replaced by glazing with cylindrical containers of water which would absorb most of the solar gain transmitted through the glazing, with night insulation and possibly selective surfaces to improve performance.

5. Control House. It was proposed that one standard Taylor Woodrow house be retained, with high levels of wall insulation, to provide a basis against which the performance of the other houses could be measured.

The thermal performance of the houses would be extensively monitored for one heating season with simulated occupancy to provide detailed energy usage and computer model verification information. The liveability of the houses should then be ascertained during a 3 year occupied period with reduced monitoring.

This project gained backing from the E.E.C. and the UK D En.

SED while being involved in other UK projects have also carried out a number of

Fig. 3 Solar Housing Development at Linford, Milton Keynes,U.K.

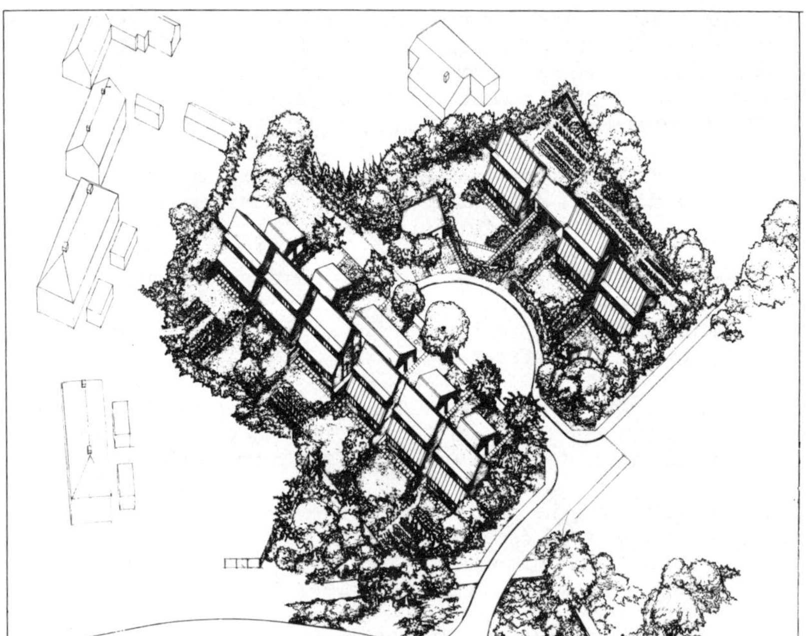

Fig. 4 Birds Eye View of the 9 Solar House Development at Linford

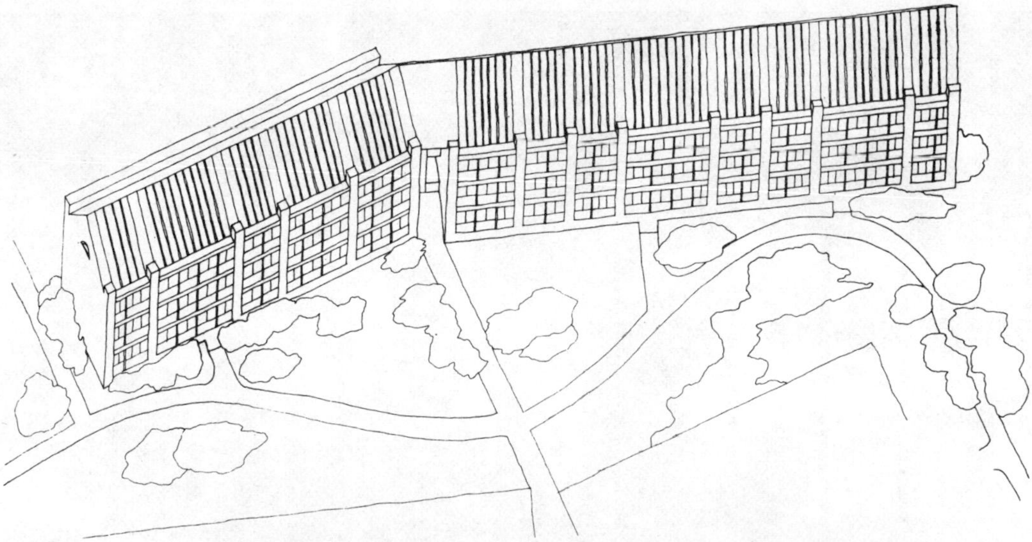

Fig. 5 Solar House Development for AGIP at Siena, Italy

Fig. 6 Model of a Solar Passive, House in Aylesbury

Fig. 7 Solar House in the Lioux Valley, France

Fig. 8 Sports Stadium in Riyadh, Saudi Arabia

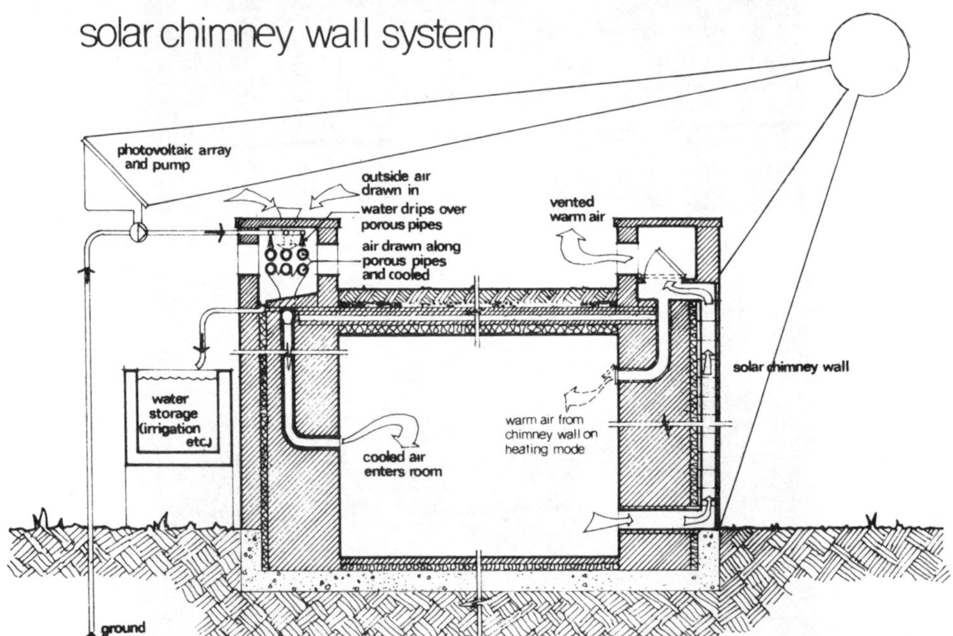

Fig. 9 Solar Cooling for a House in Marrakech, Morocco

of foreign contracts.

Two houses in France have been designed using air systems and rock stores. These two houses completed during the past year are due to be monitored by the French ministry of construction. One of these houses is situated overlooking the Lioux valley. It is planned around a broken cloister and is formed of three separate blocks, the two lower blocks being linked by a greenhouse. It is built of exposed extruded brickwork internally and externally, over a hypocaust filled with pebbles. A separate garage block has air solar collection on the roof and the heat is distributed and stored in the hypocaust for heating the rooms, swimming pool and water system.

Also the France SED, with French associates, were winners of the French national solar housing competition, HOT 5. The brief was to design a medium cost housing unit in which integrated energy saving systems add to the architectural quality of the whole without increasing the cost by more than 10%. The solution chosen was to add a "greenhouse" screen to the front of each terrace house, placed away from the main facade to provide a winter garden at lower level and balconies overlooking this at first floor level. In winter and mid season this greenhouse unit is used as a solar collector, excess warm air being taken from the top of this space to a central "thermal core" in the middle of the house which also contains the boiler and fireplace. In summer the glass screen folds away and the projecting roof and balconies form "brise-soleils" to protect the facade from the high sun. The "hybrid solar" system thus adds two distinct elements to the building, the greenhouse and the thermal radiant core.

Openings to the north are kept to a minimum to reduce thermal losses. Vertical south facing solar water panels are linked by a thermosyphon to a preheat solar hot water tank, and is planned to incorporate heat recovery on exhausted air and waste water. Overall economies in the region of 40% can be expected from the passive solar system. This design will be constructed on a site in France.

SED have also carried out architectural and energy consultancy work in Italy, for CTIP Solar and for APRE, the latter being an Energy Saving Agency set up by AGIP and Jacorossi. SED acted as architects in designing prototypical low cost low energy housing for APRE; two groups of fifty units are to be built in 1980 in Pisa and Siena. The requirement was for "popular housing" conforming to IACP regulations but saving over 30% of energy over and above the provisions of regulation 373. The first action was to orient the buildings so that their main axis ran east-west and to reduce the number of excrescences that would have increased surface to volume ratio and thus heat losses. The next action was to place as many living and bedrooms as possible due south with services and secondary bedrooms due north. The south rooms are equipped with a glazed balcony unit which can become a greenhouse in winter, whilst the glass can be pivoted away in summer so that the balcony acts as a sun shield to the flat below. The south facade is thus treated as a solar collector whilst the north is as sheltered as possible, a sheer wall with the smallest perforations allowable. In addition active solar systems are proposed to provide hot water for the flats, with collectors mounted on the pitched roofs and heat exchangers and stores in the basement.

SED have been working on a great variety of other projects in Italy ranging from greenhouse applications to Stationary Reflector/Tracking Absorber power generation.

SED has a working relationship with Studio Nervi of Rome and has been working on solar cooling systems both active and passive for some of Studio Nervi's projects in Saudi Arabia. One such project was to investigate the effect of providing a shading structure for a sports stadium at Riyadh. A cable structure consisting of nine tensile trusses was proposed to span 140 meters over the playing area of the stadium and the bleachers, with a shade net material between the cable network to provide 50% protection from the sun. SED's report analysed the microclimatic effects of the structure as well as the possibility of growing natural grass rather than laying Astroturf; and also considered the cost implications over a twenty year period.

SED's interest in passive solar cooling is continued through a project in Marrakech, Morocco. The brief here is to design and build a passive solar house with stress on the microclimate and with a secondary house in the palmeraie. The two house units are designed around a central "riad", with the generating lines on a contracting grid of intersecting squares - the Moroccan star. The construction is of the chimney wall type, creating draughts in the south facing walls by drawing air convectively up the walls through roof mounted evaporative coolers into the rooms. This chimney wall system is at present under test on site, using different materials for construction of the walls, with encouraging results.

SED has a computer facility based on an in-house computer and access via a terminal to the Ove Arup system. A comprehensive library of computer programmes is held which can for example deal with simulations of the performance of various building types under different climatic conditions, analysis of solar systems using hourly weather data, shadow plotting and economic analysis. The computer facility is invaluable in assisting SED's purpose of offering a general design and consultancy service for solar buildings and related systems.

ESTABLISHING A COMMUNITY OF SOLAR HEATED HOMES

R. Tucker

Tucker Associates, Horsepound Road, Carmel, New York 10512, USA

ABSTRACT

The experiences obtained during the establishment of a community of solar heated homes are recounted in order to assist and encourage solar community planning and construction.

These experiences include design decisions, construction management, economics of combined energy conservation and solar heating, as well as developing a marketing plan to provide customer information.

It has been found that building a solar community requires a great deal more design and planning effort than does previous residential construction. In addition, construction requires substantially more management effort, field training and testing.

Analysis of cost benefits in energy saving clearly shows that the additional effort is not only worthwhile but is virtually mandatory from any economic standpoint. The probabilities are that the major portion of future residential construction in the northeast United States will be energy conservative as well as solar, when the economic results of solar are generally known.

KEYWORDS

Solar community; site-built solar collector; energy conservative construction; economics of solar; projection of energy savings.

GENERAL DESCRIPTION OF PROJECT

The Deer Hill community is planned for 28 solar heated homes incorporating both active and passive solar heating and solar heating of domestic hot water. Deer Hill is situated in Carmel, New York, which has a latitude of 42 degrees.

Road and lot designs were arranged to provide good solar exposure from full south (180 degree azimuth) to 15 degrees west of south (195 degree azimuth). Solar tilt angles for the active panels are from 45 to 55 degrees. The panels are integrated into the roof which is at the same slope.

DESIGN DECISIONS

The project planning required coordination of architectural, solar and land design. It was decided that the houses should incorporate a site-built collector for the following reasons:

Architectural integration of solar panels and roof design.
Lower cost collectors permitted larger solar collection surface.
Substantial collection surface provided extra reliability for long term use.

The homes include many energy conservative features to reduce the energy requirements such as the following:

Passive solar heating design
Burming (earth protection from north)
Exterior insulation on foundation walls
Thick walls with 6 inches of insulation
Insulated glass windows and doors

It was decided to obtain Governmental review of overall design by submitting completed design information and calculations in a Grant Application competing for funds under the Residential Solar Demonstration Program. This required the building of a sample site-built collector for efficiency testing before and after a 30-day stagnation operating condition. The performance showed no deterioration in efficiency after 30 days of no flow solar exposure for this site-built collector design.

BUILDING PROCESS

Financing

It was anticipated that major difficulties would be encountered in obtaining financing for the construction of a solar home. This, however, was not the case as local banks were aware of the necessity of reduing home energy costs which have been increasing at an alarming rate. It is believed that the engineering review by the Government and the subsequent Grant Award helped the local banks realize the design competency of these particular solar homes.

Project Field Management

One of the surprises was the inordinate amount of management time and expense required for the proper building of these homes. This was caused by the following reasons:

(a) Special materials had to be obtained for both the solar and energy conservative features of these houses. This material could not be obtained from local building supply companies and had to be specially ordered and shipped to the site. Some examples of these are:
    - Pipe insulation for small size solar piping.
    - Exterior foundation insulation.
    - Sealants for solar tank.
    - Solar collector cover.
    - Inhibited propylene glycol heat transfer fluid.

(b) New installation procedures were required for certain parts of this construction. This, therefore, required field training of the appropriate trade for the accomplishment of the task. An example of this was the application of insulation outside the foundation which had to be wired in place and covered with a protective grout.

(c) System testing is mandatory for the solar system installation and this interrupted the construction program. The panels were pressure tested for tightness and the panel covers were tested to make sure there was no leakage from the roof. Even standard heat exchangers for the solar tank required testing prior to installation in the tank.

ECONOMICS OF SOLAR

The anticipated saving in energy by the use of solar and energy conservative home designs are substantial in the northeast United States. To provide a basis for the comparison with other locations, here is the basic heating data for one of the Deer Hill plans.

	English Units	Metric Units
House Size	1550 sq. ft.	144 sq.m
Heating Requirements	6687 deg.day F	3715 deg.day C
Inside Design Temp.	70°F	21°C
Outside Design Temp.	6°F	-14.4°C
Latitude	42 deg.	42 deg.
Av.Hot Water Use	80 gal/day	303 l/day
Collector Area	740 sq.ft.	68.75 sq.m
Tilt Angle	54°	54°
Azimuth Angle	180°	180°
Collector Efficiency		
$F_R(\tau\alpha)$	69%	69%
$F_R\ U_L$	1.06 BTU/h·ft.^{2}·F	5.17cal/h·m^2·C
Storage - Water		
Volume	1200 gal.	4542 litres

Comparative Heating Requirements

The following table gives comparative heating requirements for non-solar and solar homes in units of millions of BTU's per year.

	House Heating	Domestic Hot Water	Total Heating
Non-Solar			
Similar Size Home of Average New Construction	99.8	21.0	120.8
Solar			
Deer Hill Energy Conservative Home	52.2	21.0	73.2
Deer Hill Solar Contribution	-34.4	-19.4	-53.8
Net Energy Required	17.8	1.6	19.4
Savings by Solar			
Energy Conservative Design Savings	47.6	-	47.6
Solar Contribution	34.4	19.4	53.8
Total	82.0	19.4	101.4

Annual Savings

The current annual savings based on heating oil @ $.995 per gallon and overall oil furnace efficiency of 60% and electric heat at $10.73 per million BTU are as follows:

Annual Energy Savings		
Domestic Hot Water		
19.4 million BTU @ $10.73	=	$208.16
Heating		
82 million BTU @ $11.57	=	948.74
Annual Energy Savings		$1,156.90

Return on Investment

Current Cost of Solar System	$15,000
Current Cost of Energy Conservative Design	8,500
	$23,500
Deduct New Federal Tax Refund on Solar Installations	4,000
Net Investment	$19,500
First Year Return on Investment (1156.9/19,500) =	5.9% Return

ECONOMICS OF THE FUTURE

As can be seen at present rate of savings, the Deer Hill houses will pay for their energy conservative and solar system features at the current rate in 16.8 years (19,500/1,156.9).

This is based on the idea that the price of oil will remain constant for the next 16 years. This is, of course, an unreasonable assumption.

We can reasonably predict that the price of oil will increase each year. Stobaugh and Yergin (1979) discuss the future price of oil and the many factors which influence it. In their reference number 83, a U.S. energy official states that there is a 50 percent likelihood that oil prices will double in seven years in real terms; that is, over and above normal inflation increases. This would be a basic increase of 10% per year. An annual inflation increase of from 6 to 15% would therefore produce an annual oil price increase of from 16 to 25%.

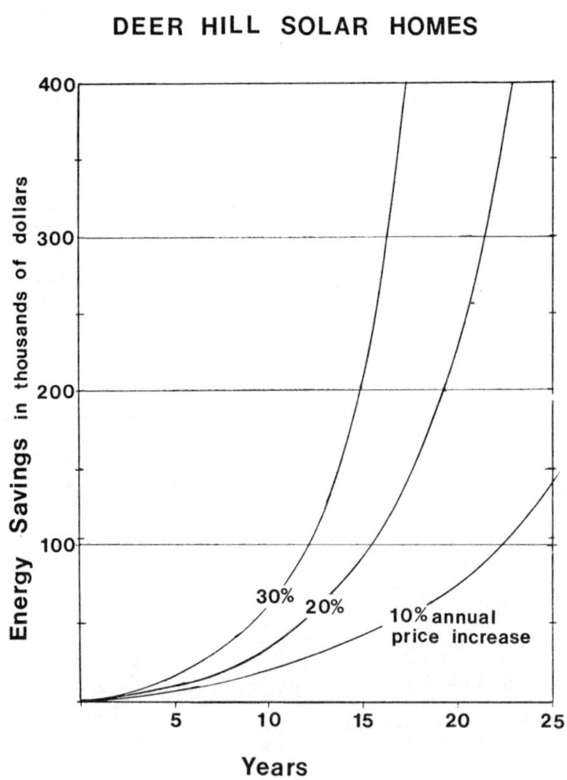

As seen from the above Energy Saving curve, the value of oil savings increases each year. A common statement about solar installations is that they will "pay for themselves in a period of 10 years". This may be true. However, as shown by the curves above, the solar system may be paid for four or five times over during the second 10-year period.

If the price of oil goes up 10% per year, the energy saving features of solar will pay for the entire price of a $110,000 house in 23 years.

If the annual increase is 20% per year, savings could be 227% of the price of the total house in 20 years, or $250,000.

If the annual price increase is 30%, savings would be 210% of the price of the total house in 15 years or $240,000.

Whether the solar system pays for itself in 16.8 years, or whether the savings are $250,000 in 15 years, is not really vital. What is important is that solar energy is economic now and those individuals who utilize solar now will benefit themselves and the society in which they live.

MARKETING PLAN

Since a sizable portion of the cost of solar homes are for savings in energy, these advantages should be properly described to prospective customers. Care should be taken in selection of the real estate brokers as new selling techniques will have to be learned.

The typical buyer of solar homes is younger, better educated, has fewer children, and the ability to make larger down payments than the average home buyer. (See Reference HUD publication Selling the Solar Home). There is currently definite enthusiasm for solar heating, yet there exists the unrealistic expectation that solar costs will decrease dramatically and that future solar costs will be lower. Since solar components are tied to building and plumbing costs, it would be more realistic to anticipate steady price increases similar to other building costs.

A marketing plan must show the benefit of conservation and solar heating and demonstrate the current reliability of this method of saving and producing home heat.

SUMMARY

Establishment of a community of solar heated homes requires a large amount of planning and project management time and expense. The rewards are the freedom from dependence on future oil and its staggering costs.

As can be seen from the energy cost saving curves, the important consideration is the reliability and longevity of the system design and construction. A system that lasts 30 years will pay for itself many times over. Reducing the initial cost of the system is not the primary design consideration.

Reliability and longevity are the key ingredients in solar design and construction. Reliability will provide home owner acceptance, and longevity will provide substantial economic benefits.

REFERENCES

Stobaugh, R. and D. Yergin (1979). ENERGY FUTURE Report of the Energy Project at the Harvard Business School. Random House Inc., New York, pp. 47 - 55 and p. 283 ref. 83.

Selling the Solar Home, U.S. Dept. of Housing and Urban Development, Solar Demonstration Program. (April 1978 HUD PDR - 296), Washington, D.C.

SOLAR ARCHITECTURE IN COLORADO ON THE THRESHOLD OF THE 80s

P. J. Tabb

Phillip Tabb Architects, 626 Mapleton Avenue, Boulder, Colorado 80302, USA

ABSTRACT

With inflation in the United States paralleling the rising cost of energy, it has been difficult to realize practical solar energy uses in building projects. Initial capital costs have been too great. Despite this fact many examples of solar architecture, especially in the Denver/Boulder area of Colorado, have successfully been designed and constructed. This paper will focus upon six (6) such projects. The types of solar systems explored will vary from solar termered and passive solar systems to large-scale active solar mechanical systems. The types of building projects will vary from single family residences to an herbal tea production facility.

KEYWORDS

Solar tempered; passive solar; active solar mechanical; crossing over all building types; projects of mass markets; hybrid systems; commercialization of solar architecture.

INTRODUCTION

The continental climate of the Denver/Boulder area, located at 40 degrees North Latitude, enjoys a great deal of incoming solar energy. In fact during the winter months, the average amount of sunshine is over 70%. However, this abundance is offset with rather cold weather. Extreme cold temperatures have fallen to below -20 degrees F (128°C). The climate of this area is characterized by over 6000 degree days (base 65 degrees F) (18.3 degrees C) per year. The summers are mild. In most instances small structures do not need air conditioning systems. The combined effects of the high percent of sunshine and the cold temperatures create a more economical environment for the development of solar space heating and domestic hot water heating.

Because of the extreme temperatures in Colorado, the use of solar energy systems must first be preceded by extensive energy conservation measures. Common conservation measures include: proper building orientation; higher R values in foundations, walls, and roofs; smaller or protected openings on the north side; the use of double glazing; the use of moveable insulation over windows; and greater attention to weatherstripping, caulking and sealing. With adequate conservation and a properly designed solar system, buildings can typically reduce their annual space heating and domestic hot water demands on conventional systems by seventy (70%) percent in Colorado.

The types of solar systems common to this area fall into the following architectural categories: solar tempered, passive solar and active solar mechanical. Brief descriptions follow:

> 1. *Solar Tempered - building projects which incorporate architectural shape and volume (form) modeling for direct gain of solar energy and for natural or free convection and distribution of warm and cool air.*
>
> 2. *Passive Solar - more extensive systems which incorporate solar collection in conjunction with solar storage without mechanical or electrical components.*
>
> 3. *Active Solar Mechanical - more sophisticated mechanical systems (usually) which incorporate solar collection, storage, distribution, auxiliary heating and control with the use of mechanical and electrical components.*

Examples of these system types can be found in the six projects featured in this paper. Many of the projects actually have each of the system types. For example, it is common for a project to have a solar tempered form, passive solar space heating system and an active hot water heating system.

PROJECT EXAMPLES

The following are examples of solar projects designed for the Denver/Boulder area. Brief descriptions and illustrations are given. It must be noted here that these six projects do not represent all of the solar work in the Colorado area, but they do demonstrate many of the principals practiced at this time.

1. Moss Family Residence is located on 65 acres in Estes Park, Colorado. Construction is expected to begin in June of 1980. The solar system designed for this house is a combination of a passive space heating system along with a thermosyphon hot water heating system. The lower level of the house is composed of a steel joist structural system filled with batt insulation and supporting an eighteen (18") inch concrete thermal mass. The south facing windows along with the moveable insulating shades are the solar collectors. The system is backed up with radiant panels. The total cost of construction is estimated at $250,000. On the following page is an isometric of the design for the house. The illustration demonstrates the south glazing, the minimum roof openings on the north side and an air-lock entry tube also on the north side. The sloped collector array is for the domestic hot water heating.

The system is designed to provide approximately eighty (80%) percent of the annual space heating and domestic hot water heating.

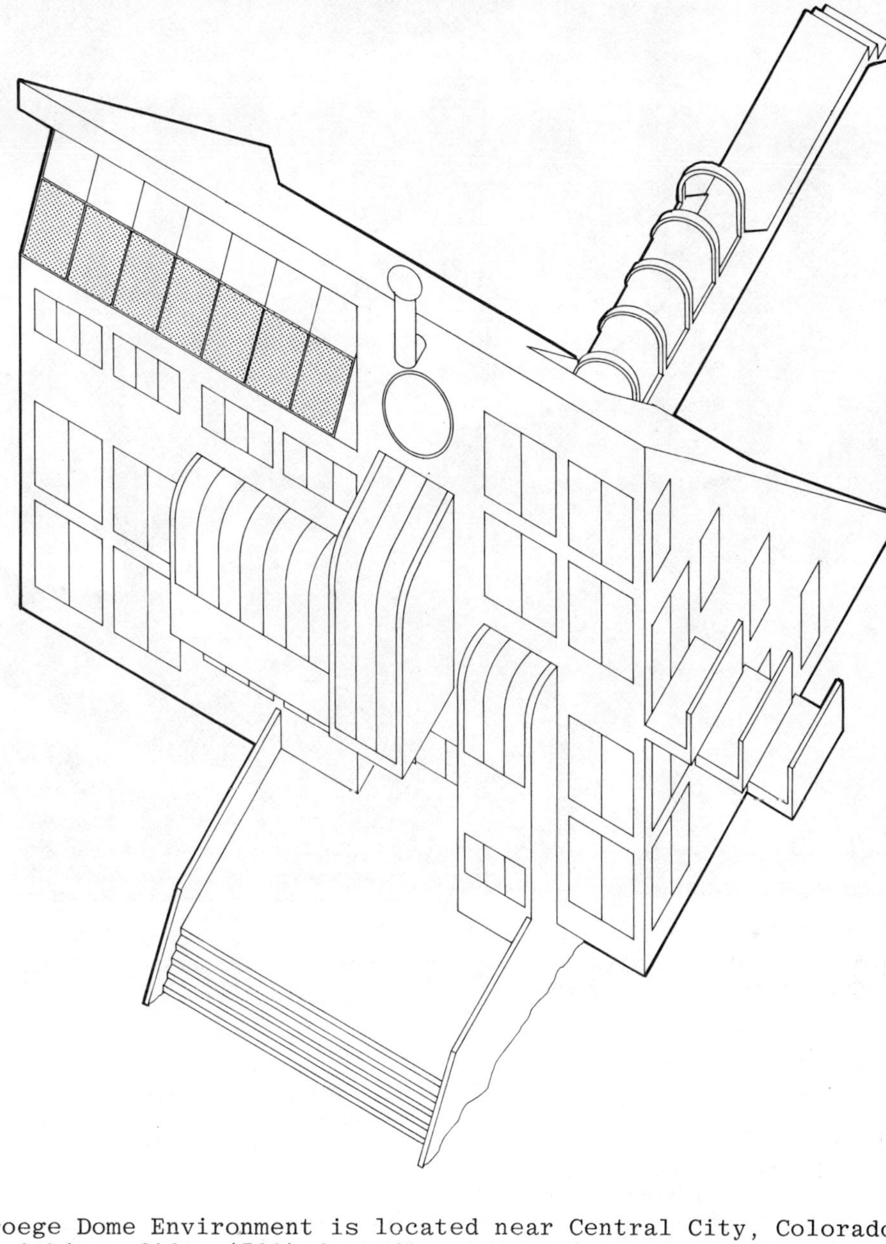

2. Droege Dome Environment is located near Central City, Colorado. Built within a fifty (50') foot diameter geodesic dome is a house which is positioned in the northern portion of the dome. The southern portion is open space behind a 350 sq. ft. area of glazing. This acts as a solar collector. At grade in this open space is an indoor pool which acts as a heat sink. The geodesic dome was built in 1972 and the house inside was built in 1975.

The photograph on the next page illustrates the house plan within the geodesic dome.

The photograph illustrates only the house. Surrounding this house is a fifty foot diameter dome.

3. The American Institute of Architects Research Corporation subcontracted for this study, 1975 by: P. Tabb. The research focused on the impact of various active solar mechanical systems upon architectural form for low-rise multi-family housing in the Denver/Boulder area. Four systems were matched with four architectural forms. The systems were: air-cooled flat plate collection systems, water-cooled flat plate systems, compound parabolic collection systems and photovoltaic systems. The isometric on the following page illustrates one of these schemes. It demonstrates the water-cooled flat plate system. Additionally, there is a passive greenhouse as a part of the overall solar collection system. This design is engineered to accomplish approximately 70% of the space heating and domestic hot water heating. The building houses ten two and three bedroom flats. Each flat is approximately 1100 sq. ft (two-bedroom) and 1350 sq. ft. (three-bedroom). The housing units are oriented along the east/west axis in order to achieve natural lighting and ventilation. The south side is dominated with collectors. The north side has minimal openings.

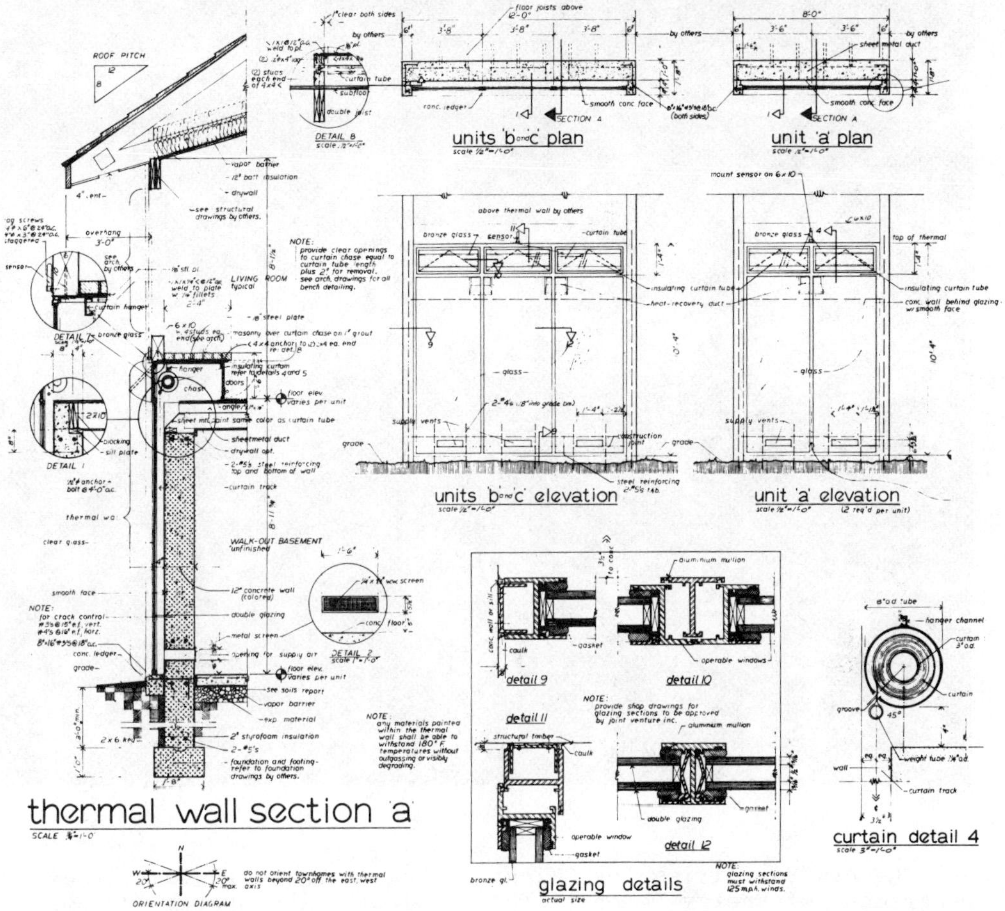

The system is composed of a twelve inch thick concrete thermal mass wall with double glazing. Between the concrete and the glazing is a moveable insulating curtain (R-10). At the top of the thermal wall is a heat recovery duct which delivers the heated air to a standard house blower which in turn distributes the warm air to the northern rooms. Each of the housing units also have an active domestic hot water heating system. The collectors are positioned on the south sloping roof.

The system is designed to deliver approximately 50% of the annual space heating and domestic hot water demands. The thermal wall collector areas per unit(s) are approximately 110 sf. and 125 sf. The lower solar participation percentage is due to the relatively small thermal wall area. Refer to the working drawing above. Two thermal wall sized are designed for two unit types. The sizes have been determined by allowable construction methods, economics and aesthetics (allowing for more visibility in the living spaces). The thermal walls are located on the basement level of the town homes.

Construction is to begin in the Spring of 1980. Occupancy is expected for the first of the town homes in the Fall of 1980.

4. The new facilities for Celestial Seasonings Herb Teas, Inc. is a Department of Energy Demonstration project. The hybrid systems of one large active collector array (9,000 sq. ft.) and two greenhouses is estimated to cost $361,000. The total project budget is over $4,500,000 in 1978.

The function of the building is for herb tea production (cleaning, cutting, mixing and bagging). The "nose-like" portion of the building which project outward to the south (please refer to the above photograph) is for administrative activities. The pyramid form is for employee meditation.

The solar systems are of standard design. The active system is a liquid-cooled flat plate system. The greenhouse system provides a preheated outside make-up air contribution.

5. Seventy two solar heated town homes are planned for the Genesee Park, Colorado project. This development is located eighteen miles west of Denver. The overall project consists of 392 housing units. Of this 392 number only 72 are being considered for solar systems. The type of solar system engineered for these town homes is a thermal wall system (similar to the "Trombe wall" system). The cost of the system per unit is approximately $7,500.

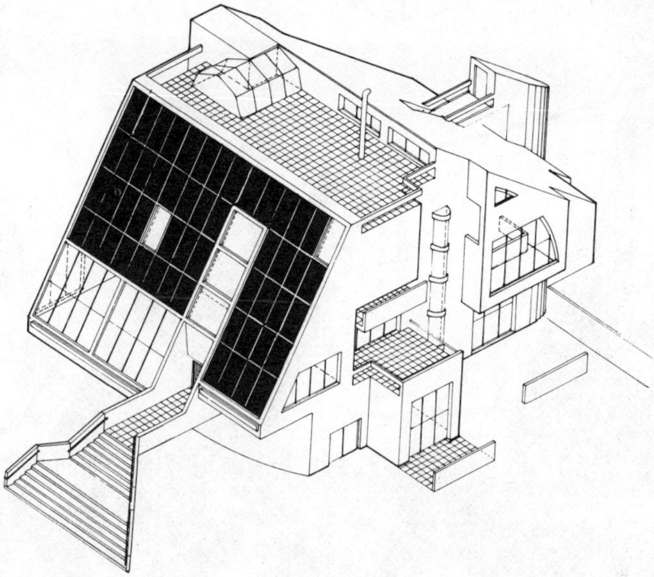

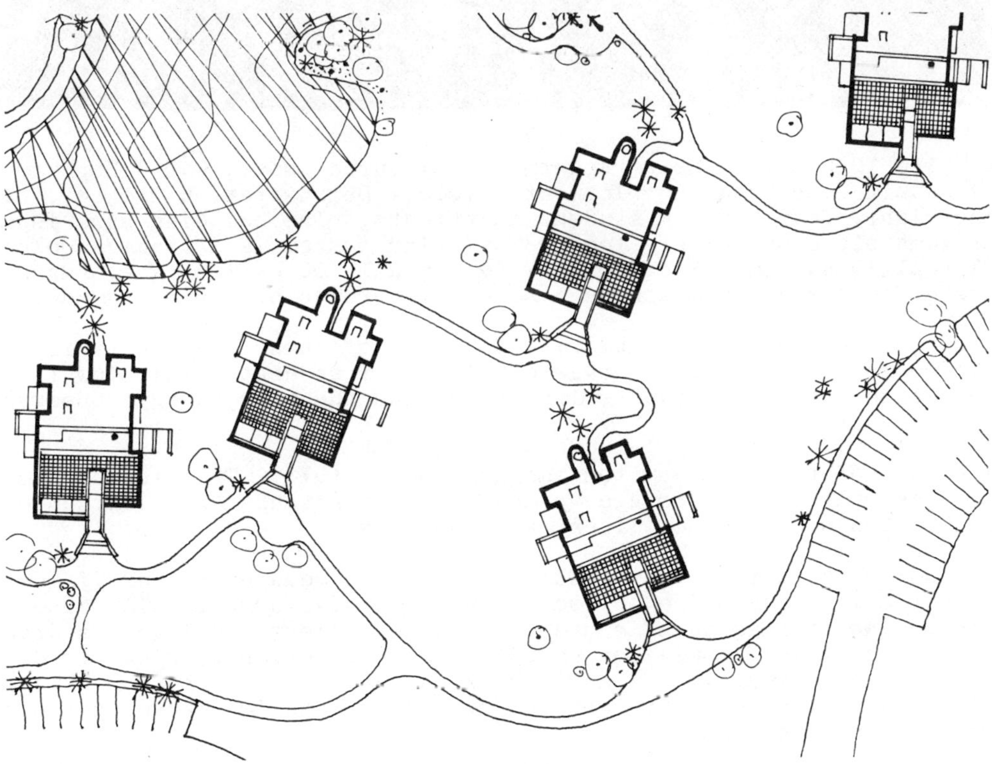

6. The 19th Street Solar Apartment, pictured above, was a Department of Housing and Urban Development, Cycle 1 Demonstration Project. It has eight, four bedroom student apartments. The building area is approximately 8,600 square feet. The solar system is a liquid-cooled flat plate system. Storage is water in a 1,700 gallon tank. Many energy conservation measures also help in reducing the energy demands.

The average monthly heating bill for each of these apartments is $11.00/mo. That is an average of $3.00 per student. This low figure id due to both the active solar system and the many energy conservation measures.

The cost of the solar system was $22,450 in 1978. This includes costs for the collectors, storage, pumps, expansion tanks, heat exchangers, collection distribution lines, and controls.

The solar collector arrays make up 864 gross square feet or 778 sq. ft. of net collection. The collectors were manufactured by KTA Corporation. The storage tank is made of 3/8" stainless steel and is wrapped with fiber-glass insulation (R-19). The collection working fluid is made of approximately 50% anti-freeze solution. The photograph above illustrates the collector arrays, overhangs, double-glazed operable windows and sun shades.

CONCLUSION

While these six projects make up a very small portion of the total solar energy projects in the Denver/Boulder area, they do demonstrate several trends. These trends with further maturing and refinement could help the overall development of solar energy for building use. The trends are as follows:

1. *Solar architectural projects are crossing over many building types. For example, there are solar heated banks, churches, post offices, animal clinics, schools, office buildings, apartment buildings, and houses.*

2. *Proven solar systems are beginning to move into projects of mass markets. This is especially true for larger housing developments and speculative office and commercial projects.*

3. *The integration of solar systems into building projects is changing from single systems to several systems including hybrid systems. The interface between both active and passive systems is proving to be beneficial.*

4. *The development of a solar industry is backing up the planning of larger-scale projects. As more and more solar projects are created, more and more solar architects and engineers are needed. In turn qualified contractors, equipment installers, persons involved in maintenance are needed. This seems to be occuring within the Denver/Boulder area.*

In conclusion, in areas of favorable solar energy it is becoming more economical to realize a variety of solar system types. In fact the commercialization of a large-scale industry is just around the corner perhaps to be fully realized in the decade of the eighties. This transition gives a welcome alternative to present day energy consumption and the resulting architectural forms. With the advent of solar architecture will hopefully be a refreshing turn for the built environment.

ACKNOWLEDGEMENT

The author of this paper would like to express his gratitude to the designers and architects of Phillip Tabb Architects.

Russel Watson, Architect
Bruce Beinfield, Desinger
Myfanwy Lloyd, Designer
Dr. Jan F. Kreider, PE

SOLAR ENERGY APPLICATIONS IN GREECE

D. Daskalakis

Heliotechnic Associates International, Fthiotidos 24, Athens 605a, Greece

ABSTRACT

This paper is a brief review of solar applications in Greece, including solar products and services being offered or established in the market today.

The first part of the paper deals with the present state of the art of solar products, namely solar collectors. The second section includes climatic data and estimates of the resulting possible energy collection. Basic economic data on currently available systems is also given. In the third section solar marketing information for Greece is presented, and the export potential of the country to the EEC countries and the international markets is explored.

KEYWORDS

Collectors; tax exemption; collector and installation prices; solar radiation; solar market; sales.

1. STATE OF THE ART

The most popular solar appliance in Greece is the thermosyphon solar water heater, various models of which are manufactured by several firms in the country. The average size of domestic installations is 2 m^2 collectors coupled to a 150 liters storage tank. Usually the same firms are also manufacturing solar collector panels for central domestic hot water (DHW) installations for dwellings or hotels.

For space heating there are several installations under construction and at least one is working at present.

Six of the most active firms manufacturing solar water heaters and collectors have established a union for the promotion of solar energy applications.

The Government realising the oil dependence of the country on its energy balance, has already taken some measures in this field and is also considering others for the near future, the most important of which are the following:

- Tax exemption for solar DHW heaters;
- low interest long term loans;
- research and development programmes;
- design and application programmes.

2. CLIMATIC AND ECONOMIC DATA FOR GREECE

Climatological conditions for the Athens area may be considered to be as near average for Greece. In Table 1 some data is presented for Athens on solar radiation and ambient air temperatures. Solar radiation values increase in the southern areas and decrease for the northern ones.

Degree days are ranging from 800 in the island of Crete to 1600 in Salonica and exceeds 2000 in several northern areas of the country.

The price of a thermosyphon solar water heater (2 m^2, 150 liters) is ranging from $550 to $800 per unit, and the cost of flat plate collectors is around $120/$m^2$ net absorbing area. The price of a medium to large size solar installation is approximately $200 - 250/$m^2$ of collector with storage tank corresponding to 75 liters/m^2.

The useful collected energy per year from such solar installations, as calculated with the f-chart method and for Athens climatological conditions is:

750 kWh/m^2 of collector for DHW with energy savings up to 80%.
260 kWh/m^2 of collector for space heating and energy savings up to 65%.

Present (January 1980) oil cost is 34 cents/kg and electricity cost is 7.5 cents/kWh, but drastic price increases are expected to be soon announced from the Government, due to the oil price increases in the international market, and the foreign currency overcharge of the external account balance of Greece.

Following a bilateral technical cooperation agreement between the Governments of Greece and F.R.G., a project is in progress for space and water heating of 500 flats in an Athens suburb, the so-called Solar Village 2 project, in which H.A.I. is participating. Preliminary calculations were carried out for an average flat of 90 m^2 area, which is well insulated (10 cm) and uses double glazing. For this appartment investment costs for various solar systems were calculated together with the relevant energy savings. The results are presented in Table 2.

The National Council for Energy and the Division of Science and Technology of the Ministry of Coordination are also funding design and demonstration projects for Solar Villages 1 and 2 in Crete. However, none of these two projects has yet progressed to the practical implementation stage.

3. SOLAR MARKET

Solar collectors first started being marketed on a systematic basis in 1975. Today a substantial market has been established following efforts by the manufacturers and the Government. Official market data is not yet available, but we have obtained good estimates from the union of solar equipment manufacturers, shown in Table 3. The sales concerned are exclusively for thermosyphon heaters and hot water installations for hotels.

In the industrial and agricultural sectors no major applications have taken place yet; however, it is now expected that a market expansion will take place following incentives by the Government for energy conservation, and the restrictions posed by energy consumption increases.

The BP oil company, which is active in solar energy applications in Greece, has initiated a research programme studying the profile of the energy consumption in Industry and Agriculture and specifically for identifying processes using low temperatures (below 100°C) hot water and air that can be met with present day solar technology of flat plate collectors.

The energy consumption of these processes corresponds to approximately 35% of the total Industrial and Agricultural energy expenditure.

Inspite of the conservative estimates of the above research programme, an annual energy saving of 60.000 tons of oil is forecasted by use of solar by 1995.

The solar space heating and cooling applications are not yet commercially developed, due to the adverse economic conditions for the investor with low present energy prices, and lack of publicly available information.

It is foreseen, however, that due to the dramatic oil price increases which are expected later this year and the Government measures, both for oil consumption reduction and the introduction of solar heating incentives that the market will soon start to develop.

Beyond the above considerations for the local solar equipment market, one should mention the possibilities of solar equipment manufacturers of this country exporting to the EEC and other countries. Greece has for years been an associate member of the EEC, and will become a full member on January 1st, 1981. As a result, all exports of industrial products to these countries will become duty free. Foreign investors are also taking advantage of the special incentives for investments, especially in specific remote areas of the country. A large German company is building now a collector manufacturing facility in northeast Greece, where the Government incentives for investments are exceptionally attractive.

Greece has also a long tradition of good commercial relations with the Arab countries, which are now becoming one of the largest markets in the world. Furthermore, for all industrial exporters, there are low interest export loans (9%) and export premiums reaching sometimes 23% of the FOB price.

TABLE 1 Meteorological Data for Athens - Greece

MONTHS	I_O kWh/m^2·day	I_H kWh/m^2·day	I_T kWh/m^2·day 53° collect angle	$K_T = \frac{I_H}{I_O}$	T_O (°C)	Degree days (ref. 18°C)
J	4.50	1.90	3.21	0,423	10,8	285
F	5.92	2.79	4.00	0,470	12,4	241
M	7.71	3.82	4.48	0,496	13,2	207
A	9.62	5.29	4.97	0,550	16,4	88
M	10.89	6.18	4.95	0,568	22,0	-
J	11.40	6.82	4.98	0,598	27,1	-
J	11.14	6.92	5.20	0,621	29,3	-
A	10.09	6.21	5.40	0,616	29,2	-
S	8.37	4.83	5.22	0,578	24,2	-
O	6.44	3.38	4.62	0,524	19,8	37
N	4.85	2.29	3.80	0,472	17,8	107
D	4.01	1.72	3.09	0,428	10,2	222
						1187

TABLE 2 Preliminary results of investment costs for Solar Village 3

No.	ENERGY SAVING SYSTEM	INVESTMENT COST PER KWH SAVED (US $)	POSSIBLE SAVING OF INITIAL LOAD (%)	FORM OF ENERGY SAVED	REMARKS
1	Insulation plus double glazing	0,53	47	Oil	
2	Passive solar systems	0,63	67	Oil	Applicable only to insolated Ap/nts
3	Active solar heating	1,00	62	Oil	Not including cost increase for low temperature heating system
4	Active solar DHW	0,34	85	Electricity	
5	Active solar interseasonal storage heating	0,90	100	Oil	Not including cost increase for low temperature heating system

TABLE 3 Sales of solar collectors in Greece

	Units	m^2
1977	4.000	8.000
1976	8.000	18.000
1979	16.000	40.000
1980 estimate	27.000	70.000

The solar water heating installation in the hotel Mediterranee (400 beds), located on the island of Kefalonia in Greece uses Thermosol modular collectors, arranged in 4 banks consisting of 48 collectors each (8 arrays x 6 collectors) totalling a net aperture area of 118 m^2, and tilted for optimum summer collection, as the hotel is mainly used during the summer.

SOME SOLAR ENERGY PROJECTS IN INDIA

S. M. Jauhri

S. M. Jauhri and Associates, 20 Central Lane, Babar Road, New Delhi, India

1. INTRODUCTION

Passive systems for both heating and cooling have a great scope of application in India where large seasonal variations in ambient conditions prevail in most parts of the country. Examples from Moghal architecture are available where use of water channels and fountains for cooling the internal spaces of palaces was quite common. The common people could hardly afford the houses for the bare requirements of shelter so that they could not think of environmental control at all.

From the middle of the 20th century and onwards, when the economic development started being felt, the need for environmental control became apparent. The predominant occupation of the people in India still continues to be agriculture, where each day the need for more production of food increases with the rapid growth of population. Therefore the use of solar devices in agriculture is very important.

The development of science and technology was hardly felt in India, when the world started to face an energy crisis. I feel that now is the right time and opportunity when we can go in the direction of solar energy utilisation straightaway without having to retrace our steps from a highly developed stage like most of the industrialised countries.

Three specific projects are presented here. Waste recycling and passive solar methods used in Poultry sheds, an experimental ecological village and an energy conserving residential apartment development.

2. PASSIVE COOLING AND HEATING OF POULTRY SHEDS NEAR DELHI

The general climatic conditions for the Delhi area are given as follows:

Summer max. temperature is 43°C with a range of 9.5°C while the winter minimum is 6°C with a range of 15°C.

Summer humidity is about 0.016 to 0.019 corresponding to Relative Humidity of 30.55%.

Fig. 1(a) General View of the Poultry Building

Fig. 1(b) The Roof Pond Installation. Note that the Canvas, Which Acts as Insulator and Radiation Reflector is in the 'Open' Position

Monsoon winds are 5 km/hr with a West direction in the morning changing to a S.E. direction in the afternoon.

The building is a poultry shed for 1500 birds and has a floor area of 450 m^2 with a height of 2.5 m. The best orientation of the long walls is per pendicular to 20° East of North so as to avoid overheating in summer. The photographs of Fig. 1 (a) and 1 (b) show a general view of the poultry and the roof pond system.

The South wall is provided with a roof projection which shades the louvred windows entirely in the summer months leaving an unshaded height of only 0.75 m above ground. The wall on the west face has a ventilated cavity.

The roof slab is on a grid pattern of R.C.C. beams and 6 cm slabs which makes 80% of the roof surface area as an effective absorbing panel. The main span beams are inverted to form roughs for a water pond and thus form a base for the moving arrangement of an insulation panel. The top surface is water proofed with bituminised glass fibre.

The floor of the building is laid with a 15 cm thick layer of rice husk, for poultry requirements, and this acts as a good insulation from the ground.

The roof water pond insulation panel is a double layer of aluminised canvas which can be collected at the end or spread out over the entire roof by a manually operated rope and pulley system.

The roof is provided with a water pond on which the insulation cover spreads leaving a 15 cm air gap during the summer days cutting off maximum solar radiation. Part of heat is also consumed by evaporation.

The radiation received inside from the walls and openings is absorbed by the water pond through the roof slab and is allowed to escape to the night sky when the insulation panel is removed. The window openings which have pivoted double asbestos panels are colour-coded so as to be opened at different times of the day or closed according to ventilation needs.

Results analysis:

(a) Temperature recorded on 24th June at 2 p.m. was 34^oC as compared to 45^oC outside.

(b) Heating in winter nights did not prove to be effective.

Remarks: It is felt that floating a polythene film on a 5 cm water layer will improve the situation in winter nights. Also the use of glass panes behind the louvres for direct heat gain, and other passive heating devices, could be introduced.

3. ECO-VILLAGE, FARIDPUR 80 km NORTH OF DELHI

The total area of land is 66 acres in the agricultural belt adjoining a prosperous village of Faridpur. This village is situated on 30^oN latitude, 77E longitude with the ambient conditions the same as those given for the poultry sheds in 2.

The aims and objectives of the Eco-Village development are stated as follows:

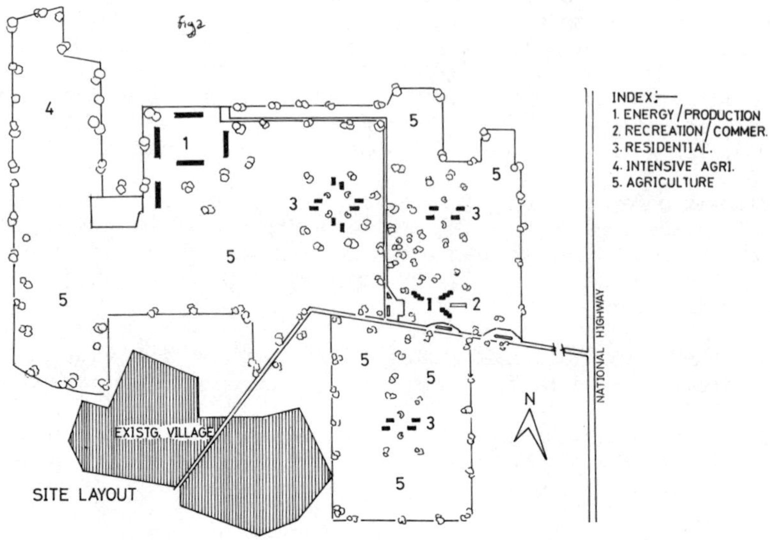

Fig. 2 Site Layout for Eco-Village at Faridpur

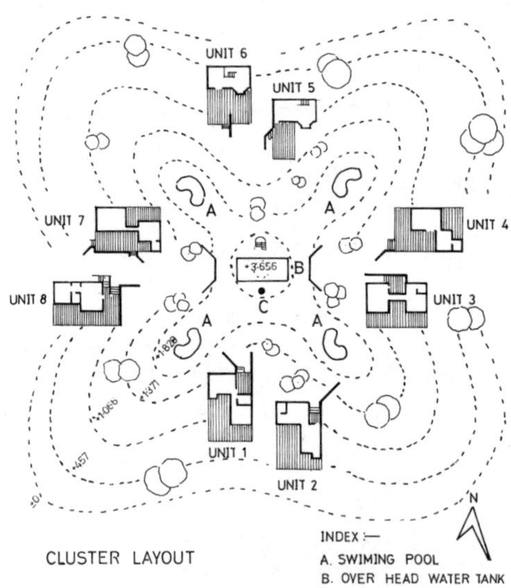

Fig. 3 Cluster Layout for Eco-Village

(a) To develop the entire land into a model village which will become self-sufficient in occupation, food production, economy and most important of all, in energy production and utilisation..

(b) To apply passive solar cooling methods and solar water heating to the small residential buildings.

(c) Other objectives concerning the use of new low-cost housing components are not the purview of this paper.

The residential zone is divided in three clusters, one of 8 dwelling units and two of 4 D.U-S. each. The layout is so designed that the common use zones 1, 2, 4 have possiblities for expansion, whereas the residential zones with agriculture have a possibility of subdividing the land in units of 4 acres each containing one or two dwellings, which is the common land-holding of an average Indian farmer.

Production of milk, eggs, fish and chicken is localised for consumption by the residents of the village and the surplus is sold to bring in revenue for maintenance and needs other than food. Similarly, agricultural produce of different kinds could be made use of.

The waste produced by food production activities is fed into 3 bio-gas plants, each placed in the middle of a housing cluster, which also receives the human waste from the houses. The gas thus produced is sufficient to provide lighting, cooking fuel, fuel for pumping out water for irrigation as well as house consumption. Three tubewells provide all the water needed out of which two are run on bio-gas engines while one is operated by a wind-mill.

Solar energy is utilised for heating water, comfort space cooling, space heating, cooking, electricity generation by photovoltaic cells, for lighting, and irrigation pumps.

A housing cluster of eight houses is located around a central green in a made-up earth mound of 3.5 m height. A water reservoir is hidden underneath and is suitably elevated to provide water to the house tops and to the fields by gravity. This mound also provides adequate privacy from all sides to each house.(See Figs. 2 + 3)

The central green will also contain the bio-gas plant in the cave like space underneath.

Small pools of water are provided at 1.5m height, to receive water from the central tank and to pass on to the fields when required.

Recreation/Commercial - This zone provides for community facilities, like a shopping centre, congregation hall & club, recreation area, and at the same time provides an outlet for sale of farm produce.

Specification for the eight houses is shown in Table 3.

4. HOUSE IN DELHI WITH PASSIVE COOLING

In flats of more than one storey passive solar cooling is tried here for the first time. The different rooms in the house are connected by a horizontal shaft (Fig. 4) near the roof level. These shafts, located in different floors, are connected to a vertical shaft rising above the roof top like a chimney for outlet of warm air from within of each room.

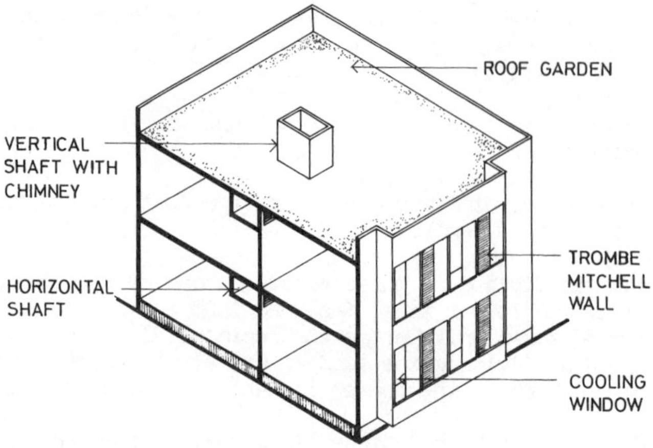

Fig. 4 Solar Heated and Cooled House for Dr (Mrs) Kusum Nangia

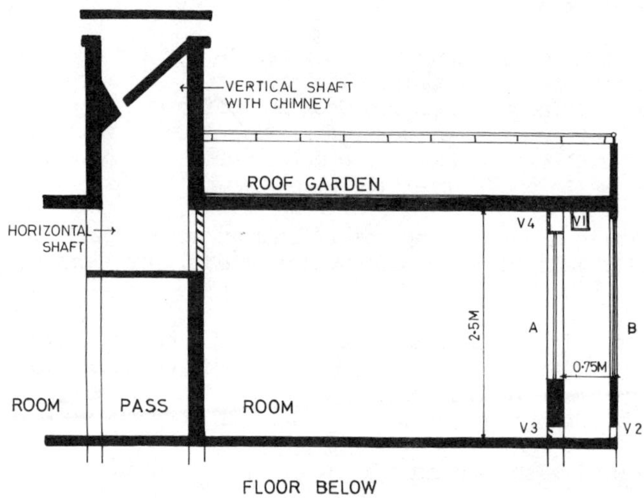

Fig. 5(a) Section of House Shown in Fig. 4

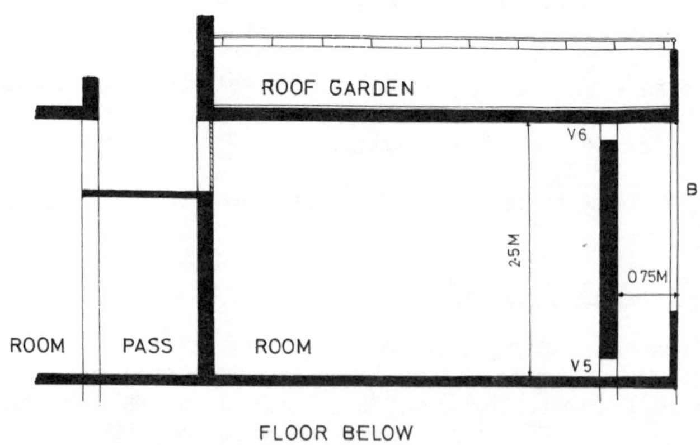

Fig. 5(b)

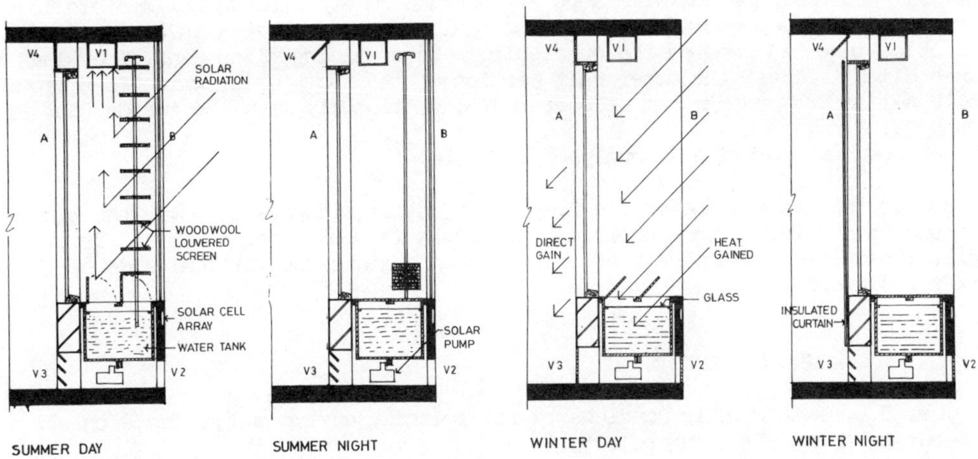

Fig. 6(a) Summer and Winter Operating Modes
6(b) of Solar Water Window

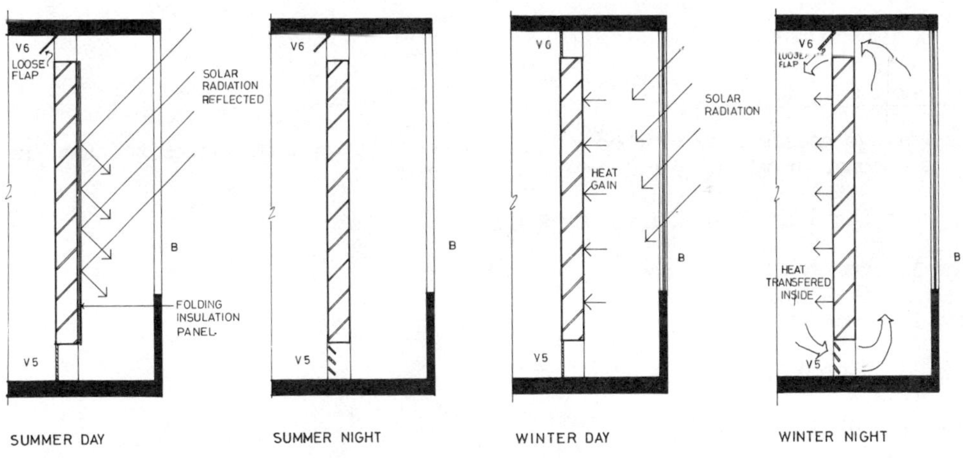

Fig. 7(a) Summer and Winter Operating Modes
7(b) of Trombe Wall

The outside vertical walls on the south are divided in two portions of equal size with a projected concrete box frame on each as shown in Fig. 5. Half portion of the wall has a window box with a water tank and an evaporative/collapsible louvred screen made of wood-wool panels. On the inside edge of the box is a hinged openable window 'A' and on the outside edge is a sliding glass pane. The glass panes slide to cover the other half portion of the south wall, which is a Trombe wall. All the components can be operated conventionally from the inside.(See Figs. 6 and 7).
The window box operation is analysed in Table 4.

In summer the glass panes of the Trombe wall slide sideways to cover the window box and hinged insulation panels of thermocole in wood panel cover the Trombe wall. These insulation panels are hinged open against the vertical side of the box for winters.

ACKNOWLEDGEMENTS

1. Dr. C.L. Gupta, TERI, Pondicherry for friendly advice and guidance on all our solar application projects.

2. Mr. Gurpreet Singh, Continental Devices Ltd, Naraina, 9A New Delhi, the sponsors of Eco-Village.

3. Mr. H. Katyal, Partner - Universal Poultry Breeding Farms - Gurgaon, New Delhi.

4. Dr. Kusum Nangia, owner of house at New Delhi.

NOTES

(1) Notes for Figure 2

The village access road offshoots the National highway and divides the land in two sections, providing easy access to the area.

The total land is divided in 5 use Zones:

Zone 1.	Energy & production	3.7%
Zone 2.	Recreation & commercial	1.5%
Zone 3.	Residential	5.0%
Zone 4.	Intensive farming	25.0%
Zone 5.	Agriculture	64.8%

(2) Notes for Table 2

(a) With the given number of animals and its resulting biogas production for pumping, water available is 30 m^3/day. With three times the number of animals the village will be self-sufficient in water pumped using biogas.

(b) The total production (437.5 kg/day) of manure will be utilised by the vegetable garden & nursery. Demand of manure for agriculture, which is now extra, will be met with the increased number of animals.

EDITOR'S NOTE

The roof thermal pond system descibed in section 2 of this paper is based on the principles developed by Harold Hay. The following references may be of use to the interested reader:

(i) H.R. HAY and J.I. YELLOT, 'A Naturally Air Conditioned Building', Mechanical Engineering, Vol. 92, No. 1, pp. 19-23.

(ii) P.W.B. NILES, 'Thermal Evaluation of a House Using a Movable Insulation Heating and Cooling System', Solar Energy, Vol. 18, No. 5, 1976.

(iii) C.L. GUPTA, 'Thermal Design Model for a Natural Airconditioning System with Application to Poultry Sheds in Hot Arid Regions', Extended Abstracts, 1975 International Solar Energy Congress and Exposition, International Solar Energy Society, 1975.

TABLE 1 Climatic Data - Eco-Village, Latitude: 30°N, Longitude: 77°E

Month	Mean Monthly Temperature (°C)		Wind Direction		Mean Monthly Wind Speed	Mean Relative Humidity	Mean Monthly Total Rain-fall	Sky Clearance Factor
	Max.	Min.	Morn.	Even.	km/h	%	mm	%
Jan.	21	6.3	W/C	NW/W	4.51	68	25	66
May	40	24	W/SE	NW/W	6.76	39	13	86
Aug.	34	25	W/SE	E/W	5.0	80	183	39
Nov.	28	10	W/C	NW/C	3.38	53	2.5	80

Note:

1. The months Jan., May, Aug. and Nov. roughly represent winter, summer, South West monsoon and autumn periods.

2. Symbol NW/W indicates that wind blows most frequency from NW and the next favoured direction is from West. The symbol C is used to denote calm.

TABLE 2 Production and Utilisation of Energy, Water, Manure for the Eco-Village of 26.4 Hectares

ZONE	FUNCTION ACTIVITY	NUMBER	AREA m^2	WATER RE-QUIREMENT m^3/day Cold	Hot	BIOGAS Production	Utilisation
Energy Zone 3.7%	Poultry	1000 birds	700	7.5		60 m^3/day	
	Dairy	50 animals	1000	3.0		20 m^3/day	50 people cooking 20 m^3/day
	Horse Breeding	25	600	8.0			
	Fish Ponds		100	1.0			
	Water Hyacinth		600				
	Circulation		6500				
Intensive Farming Zone 25.0%	Vegetable Garden		20000	25			
	Orchards	300 trees	10000	5			50 m^3/day for 5 H.P. engines for two tubewells
	Nurseries	3	2000	5			
	Greenhouses	2	500				
	Flower Gardens		4000	10			
	Fodder		120000	5			
Recreation/ Commercial Zone 1.5%	Community Centre	1	100				11 m^3/day for 40 gas lights
	Swimming Pool	1	300	4.5			
	Shops	3	50	1.0	1.0		
	Restaurant	1	200	7.5			
	Tennis Court etc.	2	600				
Residential Zone 5%	Cottages	16	1600	7.5	2.0	1.0 m^3/day	
Agricultural Zone 64.8%	Landmass		170000	90.0	3.0	81 m^3/day	81 m^3/day

TABLE 3 Faripur: Eco-Village – Specifications for an 8 Unit

Unit Nos.	Structure Description	Finishes				Energy		
		Roof (Exterior)	Walls Internal Face	Walls External Face	Floor	Cooling	Heating	Cooking + Lighting
1	Arched bamboo beams on bamboo balis and sun-dried brick walls	Thatch board	Straw board + lime plaster	Terra cotta tiles	Brick tiles glazed	Louvred windows ventilation and planting	Bio gas water heater + movable insulation	Solar stove + bio-gas
2	Pitched timber roof and cavity walls of sun-dried bricks + kiln dried bricks	Mangalore tiles	Mud plaster	Exposed bricks	Slates on rock bed	Water drip + planting	Solar panels + insulation	Solar stove + bio-gas
3	Brick vaults + cavity walls of brick and bamboos	Exposed brick painted a light colour	Bamboo painted with linseed oil	Straw board	Stone (red Agra)	Water spraying + reflection	Solar panels + solar windows	Bio-gas
4	Flat roof of bamboo reinforced cement conc. beams on bk. cols and hollow clay block walls	P.C.C. slabs	Reed matting	Lime plaster	Egg shells + lime charcoal on lime conc.	Water pond	Water pond + glass wall	Bio-gas
5	Sloping timber rafters on cavity walls of brick and slates	Slates	Rice husk board	Exposed bricks + slates	Mud with straw chippings	Water drip + ventilation	Solar panels	Solar cooker + electricity (photo-voltaics)

Cont'd ...

TABLE 3 Cont'd

Unit Nos.	Structure Description	Finishes				Energy		
		Roof (Exterior)	Walls Internal Face	Walls External Face	Floor	Cooling	Heating	Cooking + Lighting
5	Sloping roof of G.I. Pipe frame work and brick walls	Corrugated Asbestos sheets	Lime plaster + reed matting	Exposed bricks	Glazed brick tiles on rocks	Water drip and planting + exhaust fan	Solar panels + glazing	Solar cooker + electricity (photo-voltaics
7	Asbestos vaults and hollow conc. block walls	Asbestos	Lime plaster	Exposed conc.	Terrazzo on lime conc.	Water spraying + ventilation reflection	Bio-gas water heater	Bio-gas + electricity (photo - voltaics)
8	Flat roof of P.C.C. beams and brick walls	Agra stone	Rice husk board	Lime plaster painted a light col.	Agra stone (red)	Roof garden + reflection + exhaust fan	Solar panels + glass wall	Bio-gas + electricity (photo-voltaics)

TABLE 4

	Summer		Winter	
	Day	Night	Day	Night
Window 'A'	Closed	Open, if required	Closed	Closed with insulating curtain inside
Window 'B'	Closed	Partly open	Open	Open
Wood-Wool Screen	Hung up	Folded up	Removed	Removed
Water Tank	Lid open, glass panel removed	Lid closed	Lid open	Lid closed
Solar Pump	In operation	Not operating	Not operating	Not operating
Vent 1	Open	Open	Closed	Closed
Vent 2	Open	Open	Closed	Closed
Vent 3	Open	Open	Closed	Closed
Vent 4	Closed	Open	Closed	Open

SOLAR PASSIVE HOUSES IN LADAKH (N. INDIA)
The most highly situated solar buildings in the world at 4,500 m altitude

C. Stambolis, S. Watson and S. Jauhri

Heliotechnic Associates International, 5 Dryden Street, London WC2E 9NW, UK

Ladakh is one of the remotest and most ragged regions of the Himalayas. Lying north of Srinigar it borders with China on the east and with Pakistan on the west. The region is a high altitude desert with winter temperatures which fall as low as -40oC. Winter fuel-animal dung and scrub vegitation for the villagers, coke and kerosene for the inhabitants of Leh, Ladakh's capital, is scarce and precious. The combination of severe cold and smoky fuels cause much suffering and ill health, most commonly tuberculosis and eye disease. Ladakh's location (32 - 35 degrees North), altitude (3,000 - 8,500 m) and climate (an average 325 days of sun per year) create ideal circumstances for the application of solar energy.

In July 79, the Intermediate Technology Industrial Services (IT-IS), a UK based organisation which supports Appropriate Technology projects in developing countries, commissioned Heliotechnic Associates to carry out a project in Ladakh involving the survey, design and construction of two solar passive buildings. The project took place during August and September 1979 when the authors of this paper visited Leh and worked there for a total of 6 weeks. Financial support for the building work came from both IT-IS and the house owners on a 50/50 basis. Altogether three different passive systems were constructed at two selected sites. The systems included:

(a) Two rooms which employed the Solar Wall principle (Trombe-Mitchel). One room was an entirely new construction while the other was a conversion.

(b) An attached solar greenhouse.

(c) A direct gain system, adjacent to the greenhouse.

In Figs. 1 and 2 photographs are shown of the completed solar passive systems at the Karakoram and Shambha-La Hotels. The unit at Karakoram Hotel is a newly constructed room with an area of 25.2 m^2 while in Shambha-La Hotel the south facing rooms of the hotel's private wing were converted into solar passive rooms.

Fig. 3 shows a section through the room at Karakoram Hotel where the south facing Trombe solar wall is clearly shown. Table 1 also gives the basic floor and glass areas for the two systems.

Since the completion of the building work early in October all systems were reported to have performed satisfactorily. Monitoring equipment was purchased and installed with the support and co-operation of TATA's Energy Research Institute. Dr. C.L. Gupta, Director of TATA's Energy Field Research Unit (TERI), visited Leh towards the end of the building work and supervised the installation of various temperature and radiation recorders and the setting up of a general purpose weather station. Weekly monitoring reports are being completed and forwarded to TERI for analysis. Initial instrument readings from last winter indicate that the systems are a technical success. The minimum night temperature of 17°C was achieved in a solar heated room compared to 0°C in an unheated room during the same period with ambient air temperatures falling below -15°C.

The cost of building a Trombe Wall system in Ladakh is estimated at approx. 3,000 Rupees (£175). This figure, based on the average Ladakhi room size of 25 m^2, included materials and labour. Assuming many Ladakhis build their own solar wall system obtaining materials through the usual barter system, the cost of the glass alone is 2,000 Rs (£114). The price of 3,000 Rs represents approx. the same cost as 1.5 years of heating fuel; 2,000 Rs, roughly equivalent to the value of a draught animal.

TABLE 1

	Karakoram Hotel	Shambha-La Hotel
Floor Area (Solar Masonry Wall Room)	25.2 m^2 (260 ft^2)	20.07 m^2 (216 ft^2)
Floor Area (Greenhouse)	-	13.8 m^2 (141 ft^2)
Floor Area (Extension)	-	21 m^2 (226 ft^2)
Collector Glass Area (Solar Masonry Wall)	17.81 m^2 (192 ft^2)	18.6 m^2 (200 ft^2)
Glass Area (Greenhouse)	-	20.07 m^2 (216 ft^2)
Glass Area (Direct Gain)	-	4.5 m^2 (48 ft^2)
Collector Glass/ Floor Area Rations (Solar Masonry Wall)	0.7	0.9

Fig. 1 The Newly Built Solar Heated Room at Karakoram Hotel

Fig. 2 The Attached Greenhouse and Converted Solar Room at Shambha-La Hotel

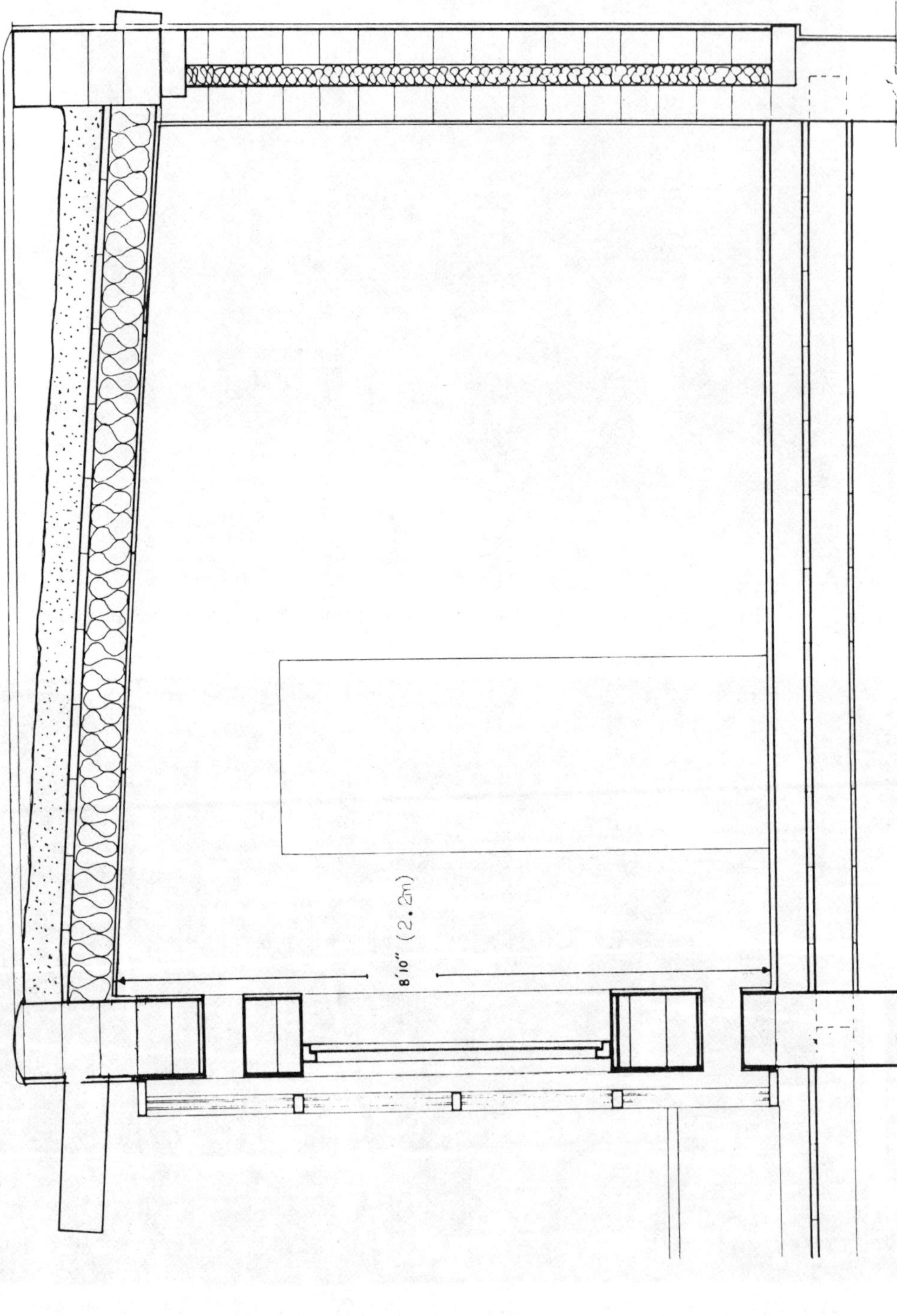

Fig. 3 Section Through the Karakaram Hotel

SOLAR POWERED SYSTEMS: SHORT REVIEW OF SOLAR POWERED FACILITIES AND COMPONENTS DEVELOPED BY DORNIER SYSTEM

K. Speidel

Neue Technologien Energietechnik, Dornier System GmbH, Postfach 1360, 7990 Friedrichshafen, Federal Republic of Germany

KEYWORDS: Water Pump, Electricity Production Unit, Cooling Process, Sea Water Desalination Facilities

CONTENTS

1. GENERAL REVIEW ABOUT SOLAR POWERED FACILITIES

1.1 Solar powered water pumps

1.2 Electricity production units

1.3 Cooling facilities

1.4 Sea water desalination facilities

1.5 General remarks about heat-pipes

2. SOLAR POWERED FACILITIES

2.1 Solar powered water pump

2.1.1 System description

2.1.2 Component description

2.1.3 Test results

3. ELECTRICITY PRODUCTION UNITS

3.1 Working principle of thermal systems

3.2 Drive designs of thermal systems

3.3 Efficiencies of thermal processes

4. SOLAR POWERED COOLING PROCESSES

5. DESALINATION FACILITIES

1. GENERAL REVIEW ABOUT SOLAR POWERED FACILITIES

Since the steep increases in the price of energy during the last few years alternative energy sources have become more attractive. Solar energy is one of these alternative energy sources, and although it is delivered free of charge, its conversion requires special components and implies investment costs. These costs become higher the lower the total efficiency of the system

Due to these aspects Dornier System has developed different solar powered facilities. The objective of those projects are:

- to gain practical experience with solar powered facilities
- to improve those systems as a result of performance evaluation.
- to study technology transfer to countries of application and enable local manufacturing of those facilities.

1.1 Solar powered water pumps

Solar powered water pumps are one of the most attractive applications in this field. They can be used for pumping drinking water supplies or for lift irrigation purposes. The development by Dornier has concentrated on water pumps for the provision of drinking water supplies. This means, that those systems must be economic in comparison to human or animal power. In case of low delivery head (5 - 6m) such an economic operation can scarcely be reached. Therefore, solar pump systems become more attractive if deep well pumping (20 - 60m) is required. For such applications thermal pumps have been developed.

The limiting factor for the total capacity is the well itself. The maximum water lifting is restricted in most cases to about 3-4 m^3 /h; otherwise the well may dry out. This means that the total capacity of those single systems is restricted to about 500 - 1000W. A 1000 W-pump system can deliver about 70 m^3/day of drinking water. That will be enough to supply a small village with 700 inhabitants, 800 heads of cattle and about 6000 m^2 of market gardening products.

Two technical systems for solar powered water pumps can be used:

- thermal systems
- photovoltaic systems

To enable a comparison between the two different systems the daily energy output of pumped water has to be compared. The total efficiency by a thermal cycle is given by

$$\eta_{tot} = \eta_{RC} \cdot \eta_{Mot} \cdot \eta_{WP} \cdot \eta_{process}$$

where the partial efficiences are:

η_{Re} = Rancine efficiency, consisting the vapor temperature and the cooling water temperature.

η_{Coll} = collector efficiency

η_{Mot} = efficiency of the motor

η_{WP} = efficiency of the water pump

$\eta_{process}$= efficiency of process, considering temperature losses and losses for the auxiliary equipment.

The total efficiency of the photovoltaic pump is given by

$$\eta_{Tot} = \eta_{Cell} \cdot \eta_{Mot} \cdot \eta_{WP} \cdot \eta_{Cel} \cdot \eta_{C_m}$$

where the partial efficiencies are:

η_{Cell} = cell efficiency

η_{Mot} = Motor efficiency

η_{WP} = efficiency of the water pump

$\eta_{C_{el}}$ = efficiency of the electrical conversion unit

η_{C_m} = efficiency of the mechanical conversion unit

The daily efficiency can be calculated as follows:

$$\eta_{tot} = \frac{h \int \dot{m} \; dt}{F \int E \; dt}$$

$\dot{m}$ = amount of pumped water at constant pump head

h = pump head

F = Collector area

E = solar radiation

The amount of daily pumped water should be related to the total amount of insolation, from sunrise to sunset, incident on the installed collector area.

The specific price for comparison of a facility can then be calculated to be:

$$P = \frac{J}{K}$$

J - investment costs of a facilities

K = daily energy gained.

This specific price enables a comparison between different types of pump system.

1.2 Electricity Production

If electric energy is desired this can be delivered too by solar thermal processes. The total efficiency can be estimated using the same method as for pumps. The only difference is that the water pump efficiency has to be replaced by the generator efficiency. An important component of the facility, which mainly influences the system design for thermal processes, is the drive engine.

Different drive engines have been designed and constructed such as:

- high speed turbines
- screw expansion engines with oil lubrication
- screw expansion engines without using lubricating oil
- turbine-gear-box-generator types
- stirling engines

The application of the engines depends mainly on the system capacity desired. (See Fig. 1).

For thermal systems flat plate solar collectors and concentrating collectors (such as parabolic troughs, paraboloids) are available to generate the heat.

The test facility reported here has a 10 kW net output capacity. A second possibility for electric power generation is the use of photovoltaic cells.

1.3 Cooling facilities

Most current cooling facilities are based on the absorption principle. Different working fluids have been considered for this application, but developments in this field have not yet been fully appraised. The types of working fluids which have proved reliable in many applications are:

- ammonia-water systems for deeper cooling processes
- lithium-bromide -water systems for climatisation processes.

1.4 Desalination facilities

Solar desalination methods using the greenhouse principle are well established. Other principles are based on multistage desalination units, using flat plate collectors or concentrating collector systems.

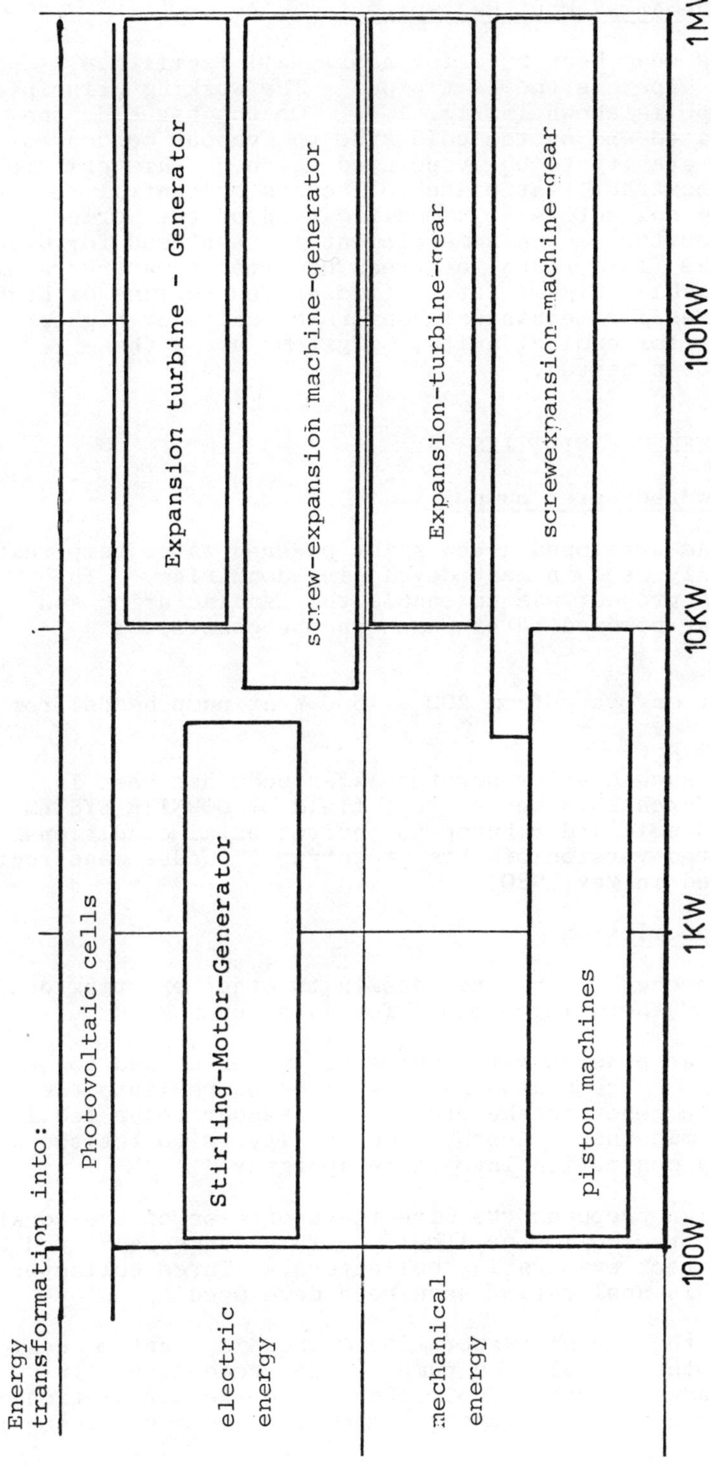

Fig. 1 Application of Different Drive Motors

1.5 General remarks about heat-pipes

One component that has been used for nearly all facilities, reported in this paper is the heat pipe. The working principle of such a heatpipe is shown in Fig. 1.1. On the hot side the liquid is evaporated and on the cold side the vapour condenses and runs back by gravity to be evaporated again. This principle can be used to construct flat plate collectors or central receivers for concentrating collectors. A modification of this principle is the film evaporator. The same element is then used for direct evaporation of the fluid (even for Freon or Ammonia-water-systems). This principle enables higher total efficiencies because of higher vapour temperatures for mechanical conversion units or higher amounts of vapour for cooling units. Systems using these principles have been tested.

2. SOLAR POWERED FACILITIES

2.1 Solar powered water pump

Dornier System has developed a new solar powered water pump that can be successfully used in many developing countries. The objective of this project was to enable the manufacturing and operation of solar powered water pumps in the country of application.

The pump capacity may vary from 200 - 1500 W at pump heads from 15 - 60 m.

The prototype of such a solar powered water pump has been in operation since March 1979 on the test field of DORNIER SYSTEM and has performed well under European environmental conditions. The second improved version of this prototype is under construction and will be tested in May 1980.

2.1.1 System Description

What is the difference of this new design to other existing designs which employ solar thermal processes for water pumps?

- Ordinary solar powered water pumps use water cooled collectors (see Fig. 2). The heated water will be pumped into the evaporator to generate the pressurized vapour to be fed into the working machine. Another possibility is to let the water flow only by convection into the evaporator.

  The new design proposes the direct evaporation of the working fluid within the collector itself. This can be achieved by different direct evaporating collectors. Three collector types for this application have been developed.

The advantage of the direct evaporation method is, that a basic energy consumer, the circulation pump, is not needed; or in comparison with a water cooled collector where the circulation is performed by convection, the vapour temperature is much higher in

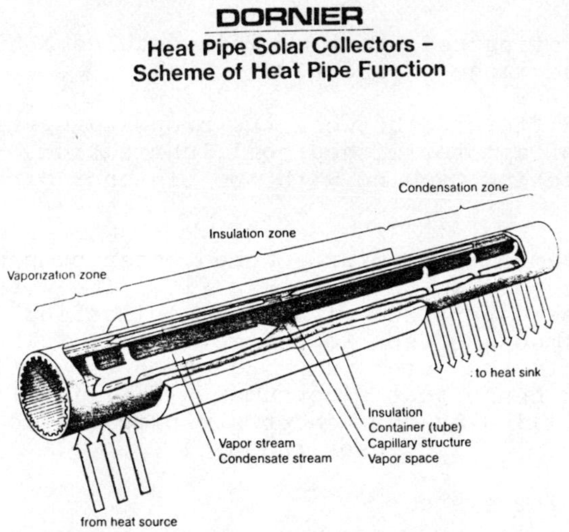

Fig. 1.1 Heat-Pipe-Function

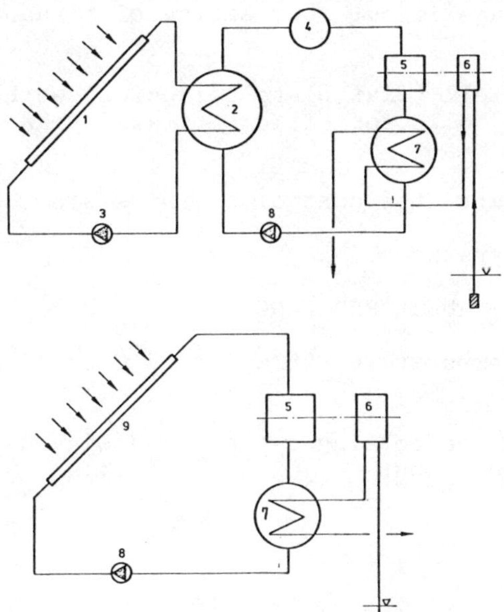

1) Directly Water Passed Collector
2) Evaporator
3) Rotate Pump for Water
4) Intermediate Storage
5) Motor
6) Water Pump
7) Condenser
8) Condense Pump
9) Direct Evaporator

Fig. 2 Functional Diagram of a Solar Pump With Indirect or Direct Evaporation

the direct evaporating collector. This enables a higher system performance in the range of about 20 - 30 %.

The second main difference between the proposed system to ordinary ones is the use of a motor without oil lubrication. The developed motor is a slow acting machine with two pistons, directly coupled to a flywheel.

A schematic diagram of the solar powered water pump with direct evaporating collectors is shown in Fig. 1. The working fluid (Freon 11) will be vaporized within the collecting field (9) and fed over an intermediate tank to the motor (5). The motor is directly coupled to the water pump (6) which pumps the water to the condenser (7) and hence to the consumer. The liquified working media will be refilled by the condensate pump (8) to the evaporator (collector) again. This closes the cycle.

Specification of the system

- Working without any additional energy source
- motor, working without lubricating oil; capacity range about 200 - 1500 W;
- collecting field (flat plate collector) working as an evaporator; collector efficiency at designed point about 40 % (90° C)
- automatic start/stop control of the system.
- Working temperatures:

  vapour input, about 85° - 90° C

  Condensate-temperature, 25°C (if 20°C cooling water is available).

All components of the working cycle are designed for a capacity of 1 kW net hydraulic output.

2.1.2 Component Description

The main components of the system are:

- collecting field, designed for direct evaporation
- motor,
- water pump
- auxiliary equipment

Collecting field

The designed collector is a flat plate collector with two glass covers. Each collector module (See Fig. 3) consists of 6 heat-pipes designed as individuals flat plate collectors. On the top of heat pipe the working fluid will be vapourized in a separate heat-exchanger. The vapour will then feed to the storage tank. Each module has a length of 5,6 m and a width of 1,25m. Two additional alternative versions, an east-west orientated collector module (see Fig. 4) and a high efficient film evaporator has been developed.

Motor

The especially designed solar motor is a double-acting piston machine as shown in Figure 5. This machine is directly coupled over a flywheel to the water pump. This engine is working with low piston speed and was designed to operate without lubricating oil.

Water pump

A piston water pump was chosen to enable operation at higher efficiencies. The internal efficiency is about 82%. This pump is directly connected to the motor.

Auxiliary equipment

In order to drive this closed cycle automatically the following auxiliary equipments are foreseen:

- automatic start-stop valves
- condenser and precooler
- condensate pump
- connecting pipes and different valves

The total system exclusive of the evaporator, which is integrated in the collector area, is mounted on one frame.

Fig. 6 shows the total view of the system with a large area on the top for testing purposes. On this platform the motor, the vapour storage tank, the auxiliary equipment and the condenser are mounted. (see Fig. 7).

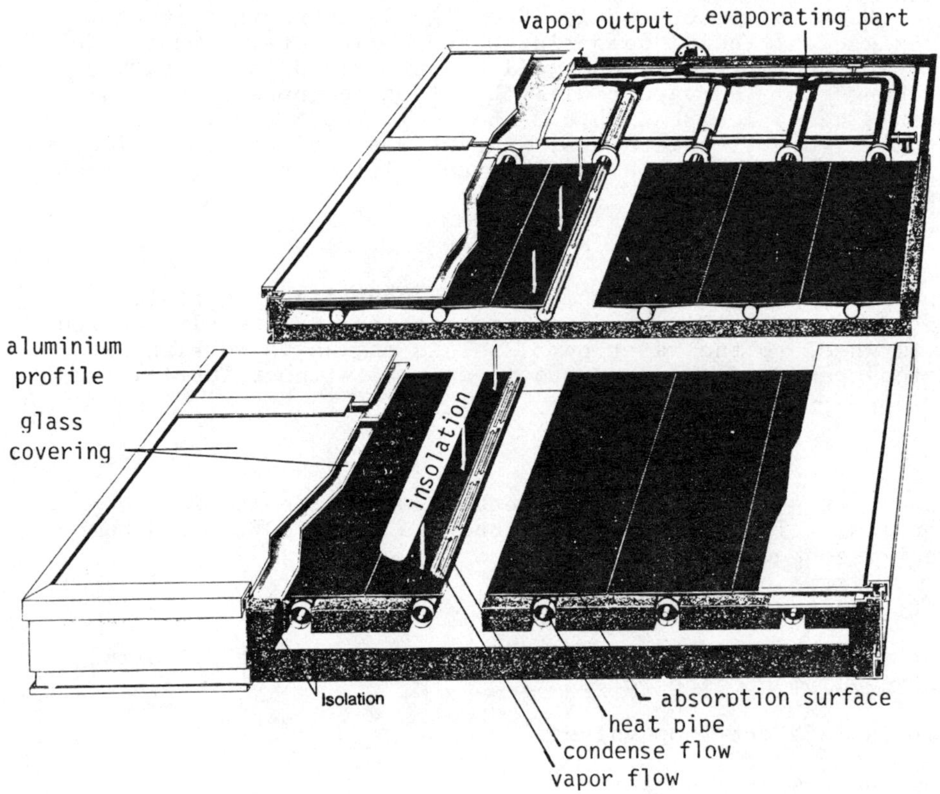

Fig. 3 Evaporator Design

Fig. 4

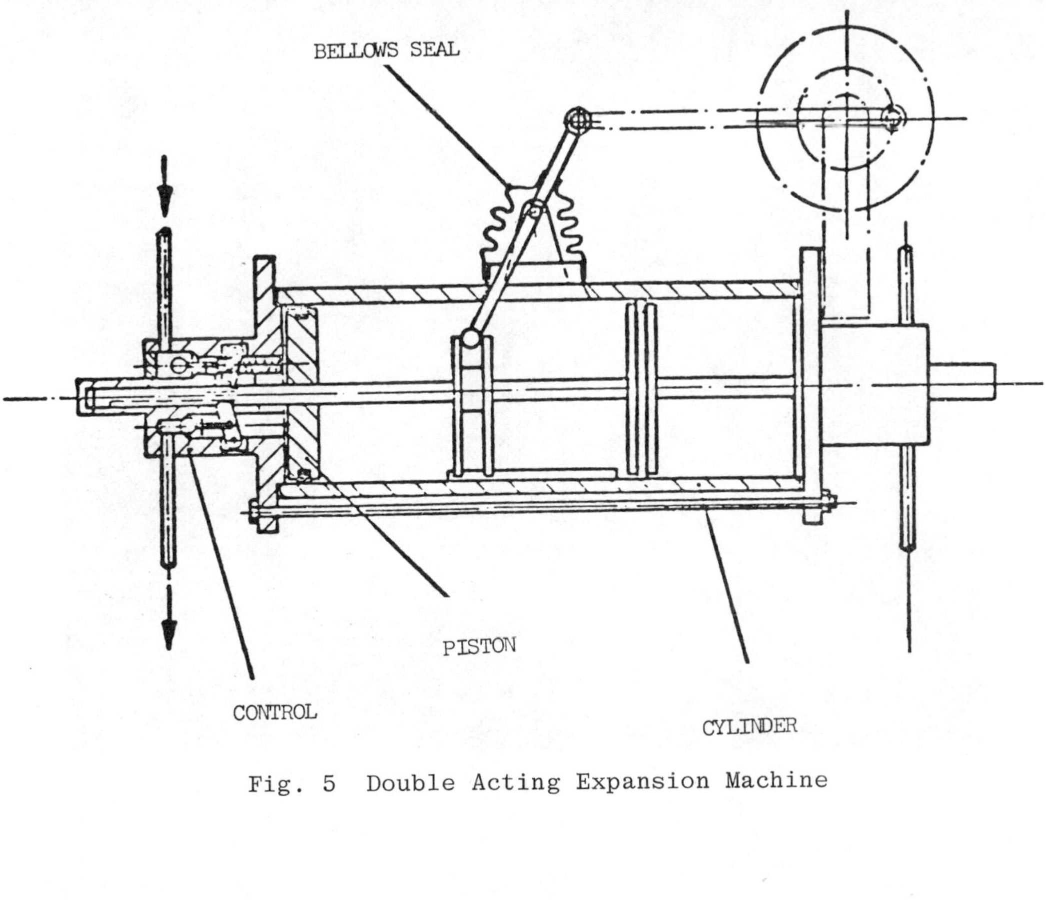

Fig. 5 Double Acting Expansion Machine

Fig. 6 General View of Solar Water Pump Installation

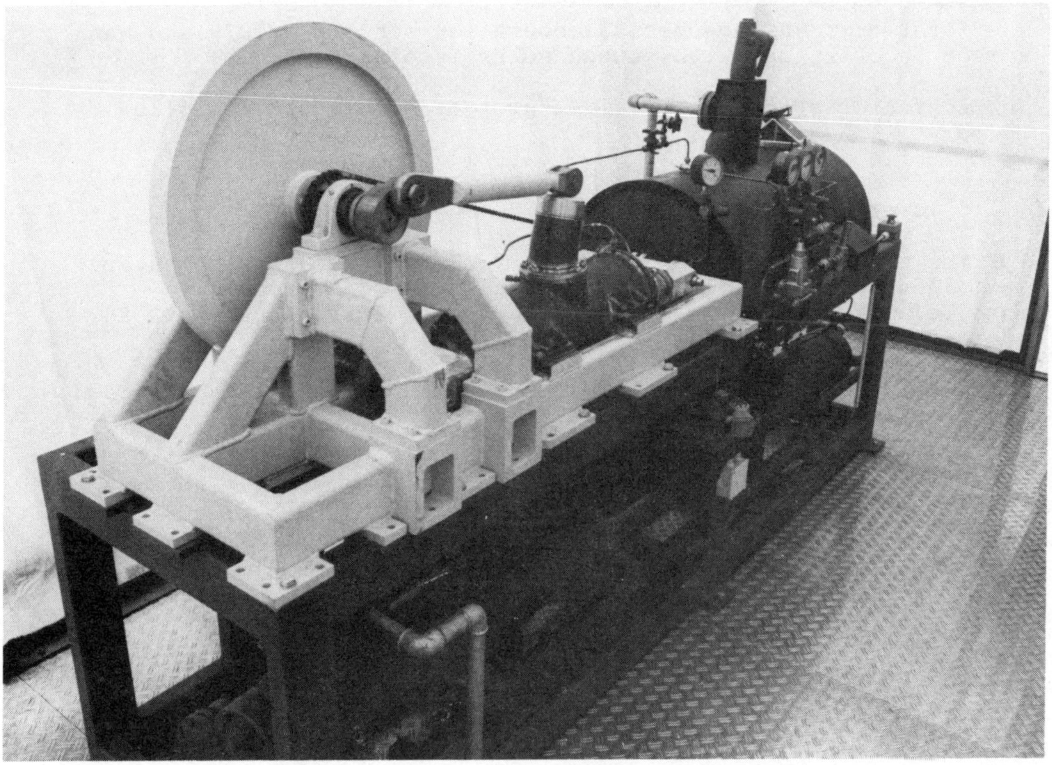

Fig. 7 Freon Cycle With Motor

2.1.3 Test Results

The first unit has now been in operation since May 1979 and has worked on every day with enough solar insolation.

The characteristics of the first prototype unit are as follows:

collector field: 21 m^2

motor capacity: 0, 1 - 1, 2kW

For the motor capacity of about 1, 2 kW (maximum) the momentary installed collector field is too small. In consequence the total efficiency for this test unit is lower than that with an adapted collector field.

With this configuration the following results have been obtained (measurements with environmental conditions in Germany)

- Total system efficiency (maximum) 2.0%

  (defined as : $\frac{\text{energy of pumped water}}{\text{solar insolation on collector field installed}}$)

- average efficiency during operating time, about 1,8%
- pump head 25m (other pump heads possible and may go until about 150 m)
- cooling water temperature: 21°C
- ambient temperature 18 - 25°C
- insolation between 450 - 850 (W/m^2)

These test results have been used as a base in order to calculate the solar pump performance of this system for ambient conditions in hot areas e.g. in Cairo, Egypt in July. (The insolation values are measured values for July 1978). The transformed values (from measurement results in Germany related to the climatic conditions of Cairo)have taken into consideration the following factors:

- the prevailing higher ambient temperatures (25°C max in Germany to 40°C in Cairo)
- the higher total efficiency when the collector area is adapted to the motor capacity, which result in relatively lower losses and higher temperatures within the evaporator (90°C instead of 75°).
- optimization of the system.
- the angle between collector and the horizontal is 30° instead of an optimum angle of about 60°C for Germany

Without taking into account the above factors and without considering the better ambient conditions, the following results have been obtained:

- total daily energy output about 8 kWh/d which corresponds to 115 000 Liter/day, from 25 meters manometric head

The following system characteristic can be demonstrated:

- The system starts operating after a preheating period of about 1,5 hours, where the insolation increases from zero to 500 W/m^2 (average insolation about 250 W/m^2).

- The maximum system capacity is reached about 0, 5 hour later than the maximum insolation (energy storage within the system)

- the total efficiency in the evening is higher than in the morning. That depends on the heat capacity of the system and higher environmental temperatures in the evening (a certain amount of the solar energy from the preheating period is transferred to drive thesystem late in the evening); therefore the system stops if the insolation on the collector field is lower than about 450 W/m^2 and starts at about 500 W/m^2 insolation.

- The daily working period of the system will be about 8 hours.

In our calculations a collector field of $60m^2$ has been assumed. At design point the system capacity is about 1 kW hydraulic power output at 800 W/m^2 insolation and the system performance factor on average, as defined above, is about 1, 8 %.
Fig. 8 illustrates the daily performance of the system under consideration.

3. ELECTRICITY PRODUCTION

3.1 Working principle of thermal systems

Fig. 13 shows a simplified flow scheme of a thermal system using flat plate and concentrating collectors. The flat plate collectors will be used for preheating the working fluid and the concentrating collectors are foreseen for evaporation of the working fluid. The vapour will be stored before being fed to the turbine. The output of the turbine will be cooled so as to obtain the required energy for preheating the liquid. After condensing the liquid will be fed over the above mentioned recuperator to the preheater and the evaporator. A storage tank is foreseen to enable electricity production during night. Fig. 14 represents a total view of the facility which has been installed in Cairo, Egypt.

The working principle of a solar stirling engine is shown in Fig.15.

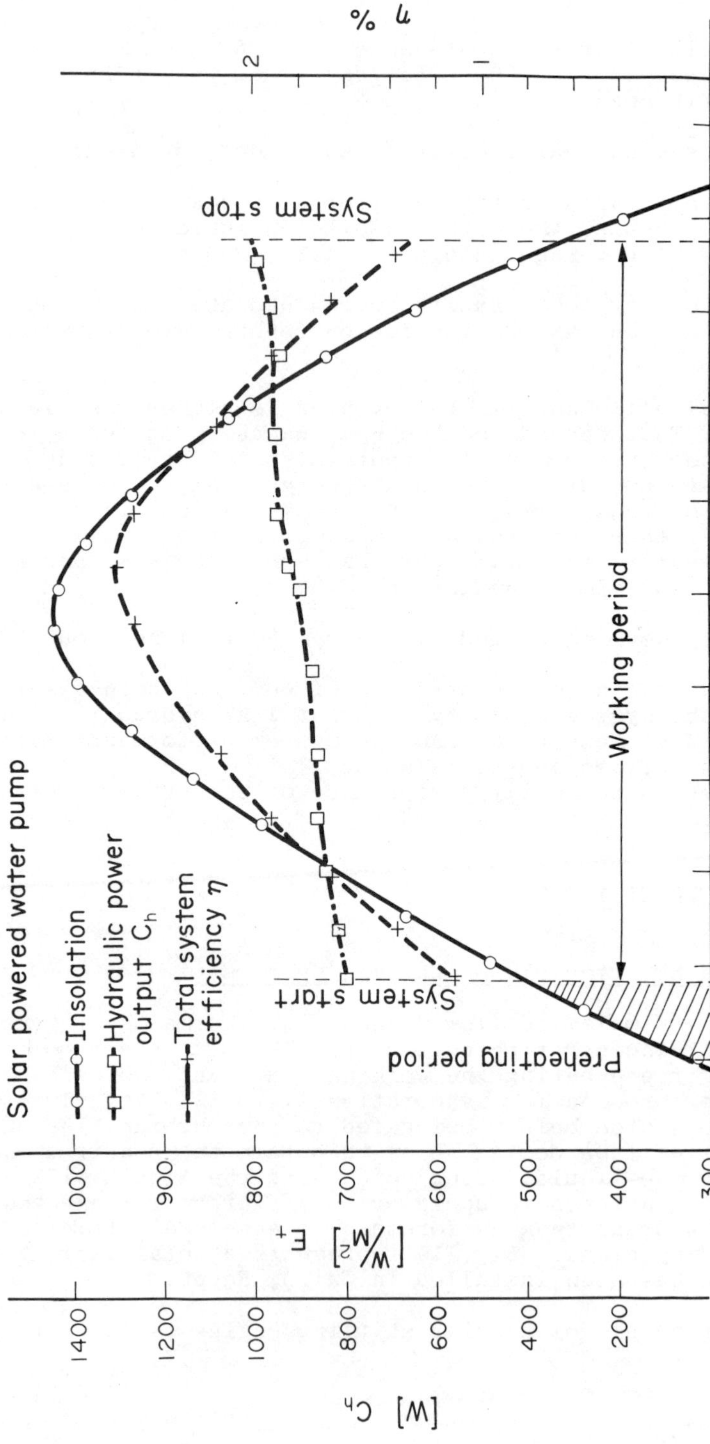

Fig. 8 Transformed Measurement Results to Cairo Insolation Conditions in July

Fig. 9 Screw Expansion Machine With Coupled Generator

Fig. 10 Turbine-Generator

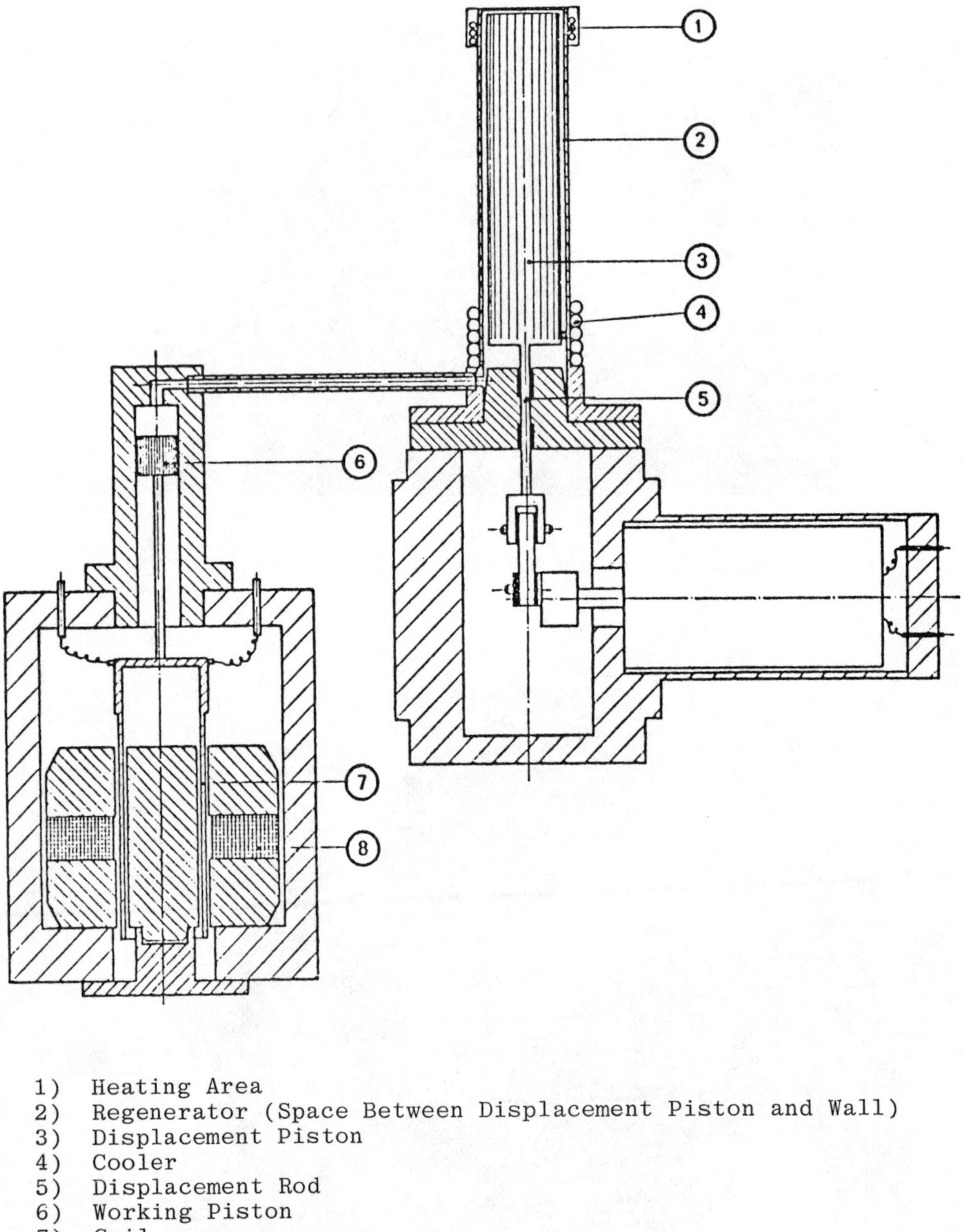

1) Heating Area
2) Regenerator (Space Between Displacement Piston and Wall)
3) Displacement Piston
4) Cooler
5) Displacement Rod
6) Working Piston
7) Coil
8) Permanent Magnet

Fig. 11 Free Piston Stirling Machine

	component efficiencies	η_{Coll} (%)	η_{RC} (%)	η_{MOT} (%)	η_{Gen} (%)	η_{P} (%)	η_{WP} (%)	η_{Tot} (%)
Small scale pump Systems 1 - 5 kW	present state	50	13 (11)	35	-	80 (58)	80	(0,8) 1,5
	developmental possibilities	55	20	70	-	85	80	5,0
Stirling-Motor-Generator 0,1 up to 8 kWel	state of the technic	35	15			75.	-	4,0
	developmental possibilities	40	25 - 30			80	-	8,5
Solar-electric Systems ca.5 - 30 kW	state of the technic	50	15	75	85	65	80	2,5
	developmental possibilities	60	40	80	90	75	80	10,3

Fig. 12. Total and Partial Efficiencies of Different Thermal Processes

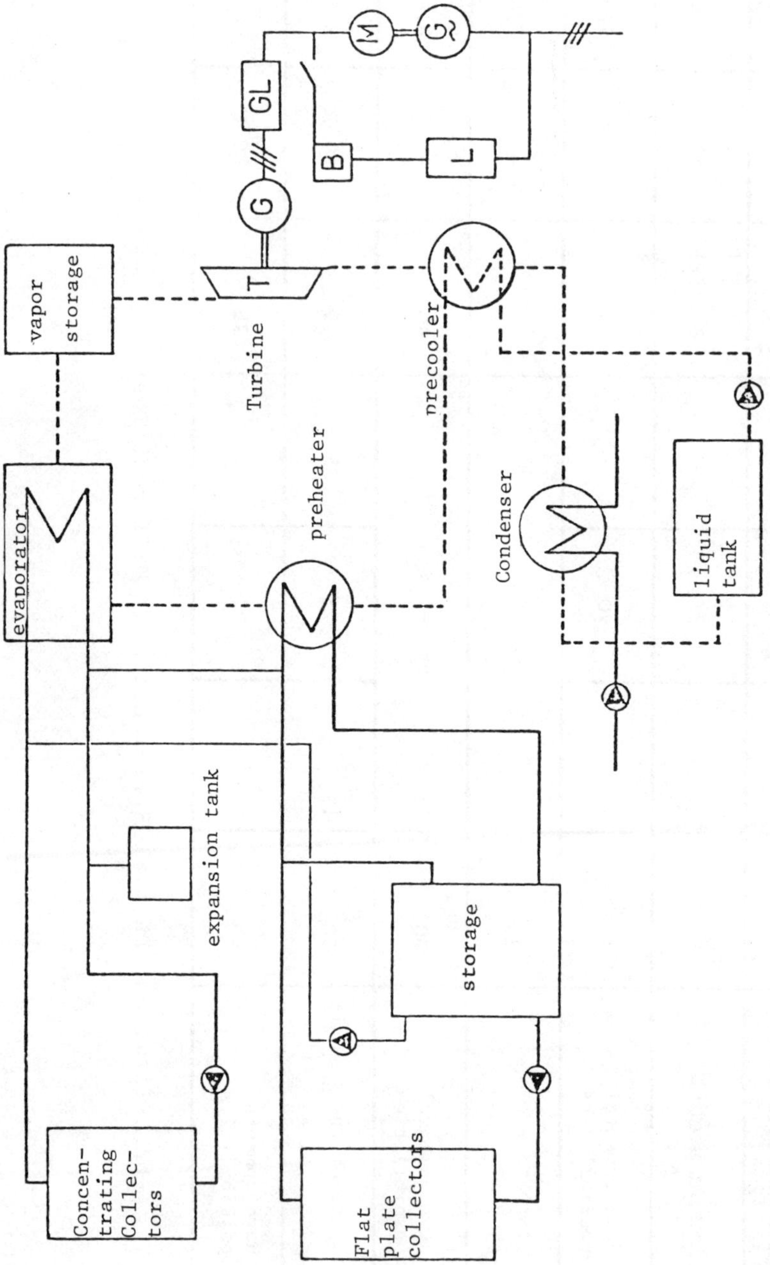

Fig. 13 Flow Scheme of a Thermal System

Fig. 14
General View of the Pilot Power Generation
Facility in Cairo

3.2 Drive designs for thermal systems

The required capacity is an important limiting factor for the use of different drive designs (Fig. 1):

- piston engines are mainly used for low capacity ranges from about 100 W - to 10 kW. Those units will be mostly used for facilities, where mechanical energy is needed (water pump etc.).
- The application of screw expansion machines begins at about 4 or 5 kW and has a wide range of capacity ranges.
- Expansion turbines will be used for capacity ranges beginning form 10kW (high speed turbines).
- Stirling motors for solar applications, for direct conversion into electricity have been developed for capacity ranges from 100 W to 6 or 7 kW.

Screw expansion engines and turbines can be used for low and high temperatures (from 80°C to 300°C). The limiting factor for high temperatures is not the engine , but the liquid to be used for energy transportation. The use of free piston stirling engines is restricted to high temperature applications. The minimum temperature is about 300°C.

In the following figures, 9 to 11, a screw expansion engine with coupled generators, a turbine and a free piston stirling engine is shown. (The free piston stirling engines have been developed by Philips, Eindhoven).

3.3 Efficiencies of solar thermal processes

As shown in Fig. 12 which tabulates the total and partial efficiencies of different thermal processes these processes have a great developmental possiblility. For small scaled water pumps 5% efficiency could be reached. That means a collector array of about 25 m^2/kW, which represents the main total costs of these systems. For those facilities where flat plate collectors can be used at those high efficiencies, the attainable total efficiency for solar Stirling units will be about 8, 5 %, which means a specific collector array of about 18 m^2/kW, assuming 80 % of direct radiation (such systems work only with concentrating collectors). Solar powered electric conversion units will have a maximum efficiency of about 10 % (for 5 - 30 kW-units). The specific collector area will then be in the range of about 15 m^2/kW.

4. SOLAR POWERED COOLING PROCESSES

One of the most important cooling systems for deep cooling purposes, is the ammonia-water system. The principle of the absorption loop is shown in Fig. 16. In comparison to other facilities this cooling unit uses the collector itself as a generator. After rectification the generated ammonia is fed to the evaporator (cooling house).

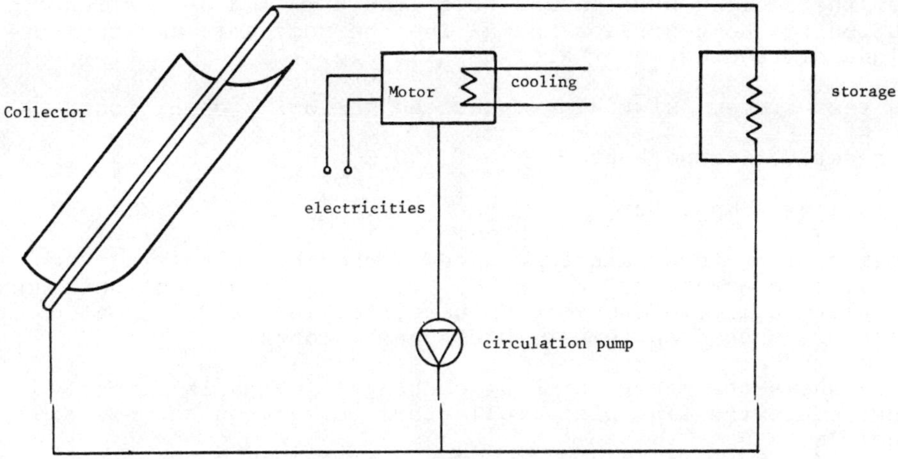

Fig. 15 Solar-Stirling Unit (Schematically)

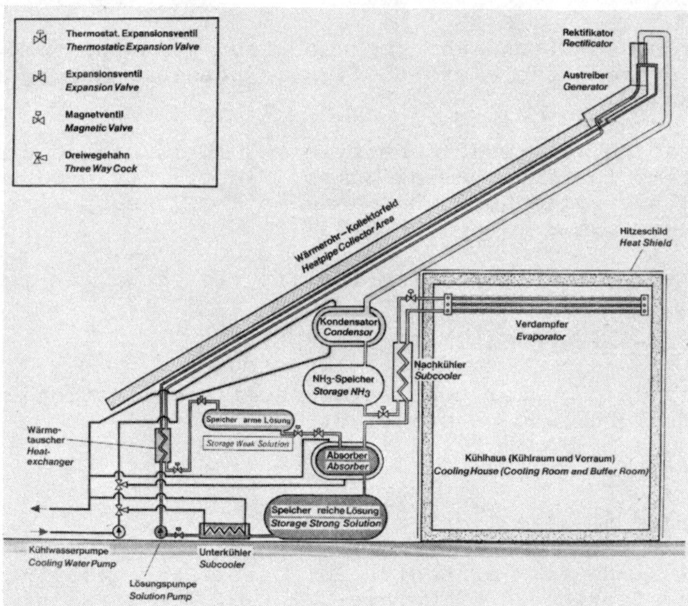

Fig. 16 Solar Cooling-Scheme of Function of an Absorption Cooling Plant

The absorber combines the weak solution and the expanded ammonia after the cooling process and this liquid is fed as a strong solution to the generator again. The thermodynamic working conditions of the test unit are shown in Fig. 17.

This test system is working under the following conditions:

- generator temperature 90°C
- cooling temperature 1°C

Deeper cooling temperatures can be reached by increasing the generator temperature (18 - 20°C). For this first test unit normal flat plate collectors have been used. For selectively coated collectors such deep temperatures can be reached.

Fig 18 shows the generator-heat exchanger, which is directly connected to the flat plate collector, working on the heatpipe principle.

This test facility has the following technical properties:

- flat plate collector area: 20 m^2
- cooling capacities of 3kW at 750 W/m^2 insolation
- Collector efficiency of about 40 % at 100°C generator temperature
- the strong solution, the weak solution and the ammonia can be stored; therefore cooling is possible during the whole day.

At the moment an additional energy of 200 W is needed during operating time for the cooling pumps. Electricity produced by means of photovoltaic cells is used to enable a self-supporting operation of the system.

An overall view of this testing facility is shown in Fig. 19.

5. SEA-WATER-DESALINATION-FACILITIES

Different types of solar powered seawater desalination units have been designed and constructed:

- solar stills
- multistage desalination units.

Some technical data is presented for multistage desalination units, desinged and fabricated by Dornier System. The first system is employing heat-pipes at two stages (see Fig. 20).

The first stage consists of a collector field, equipped with heat-pipes (see Fig 21). The heat is transferred to the above installed basin, which is filled with sea water. The generated vapour condenses on a cooler, which is the hot side of the second

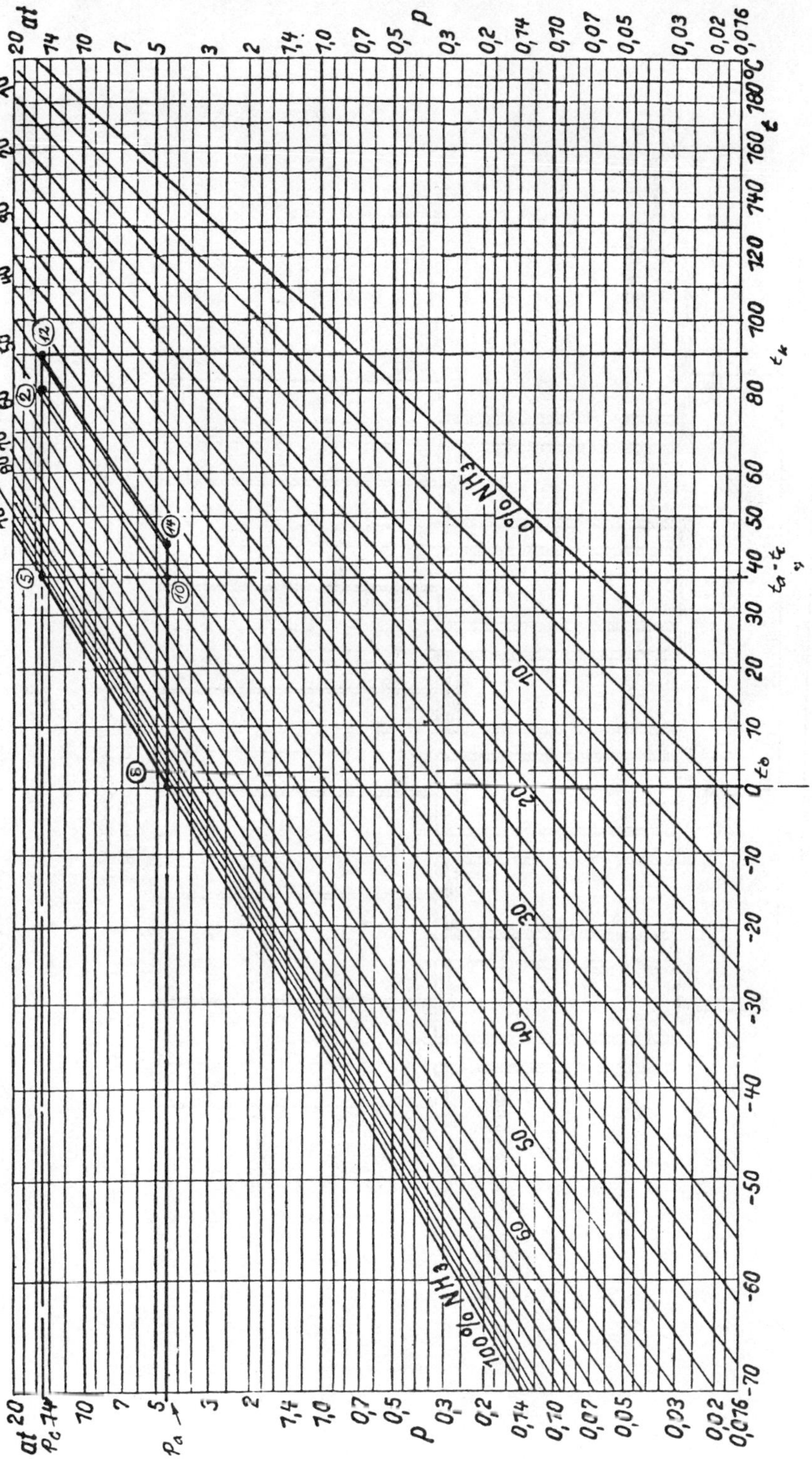

Fig. 17 Thermodynamic Working Conditions for Cooling Cycle

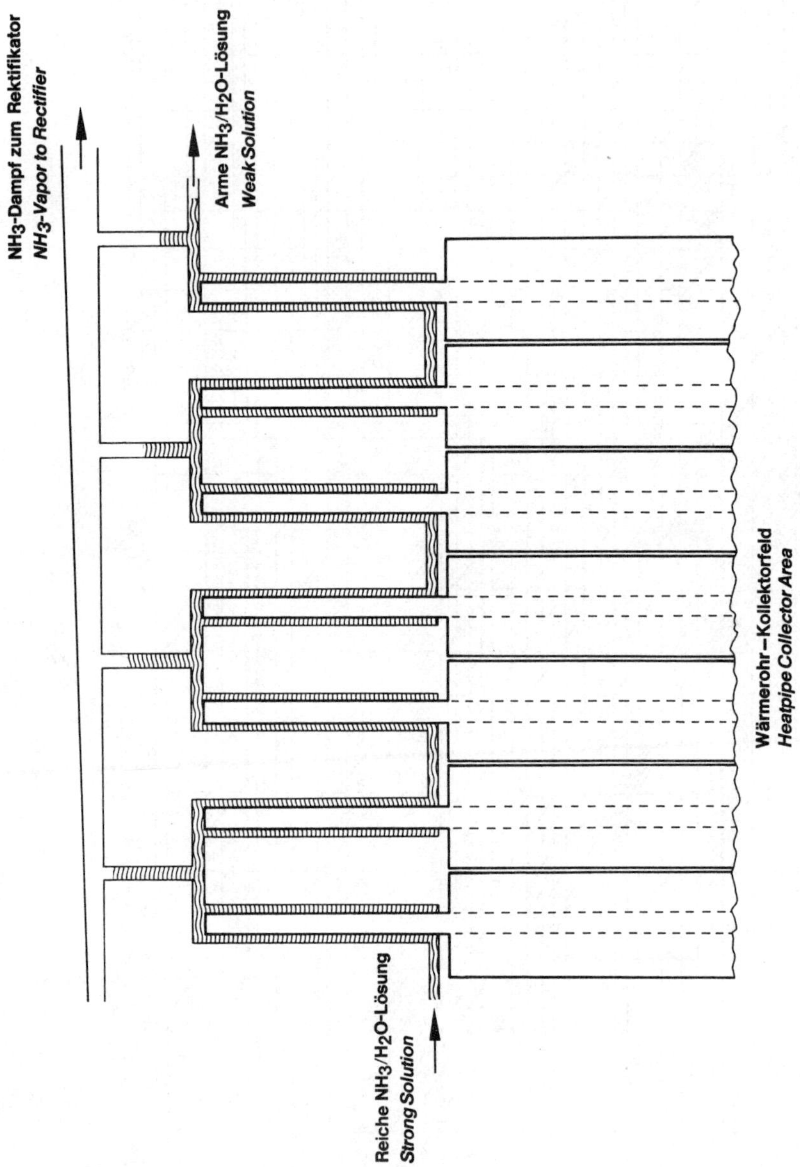

Fig. 18 Generator-Heat Exchanger (Series Connection)

Fig. 19 Total View of the Cooling System at Cairo

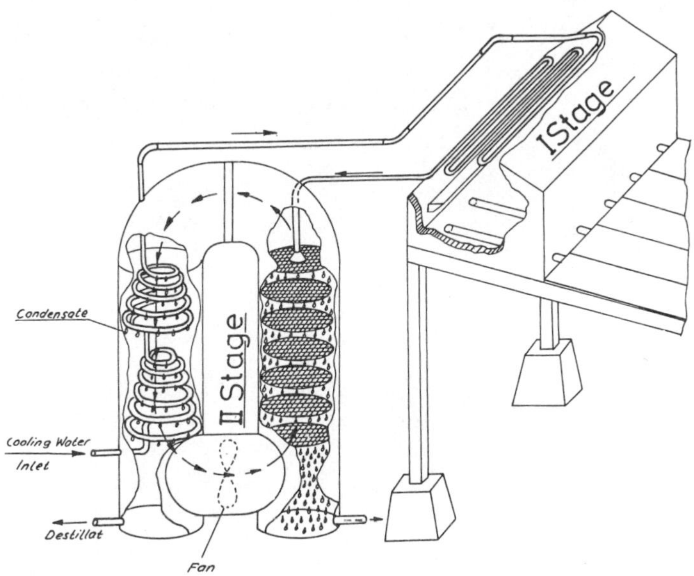

Fig. 20 Solar Desalination Unit, Using Heat Pipes

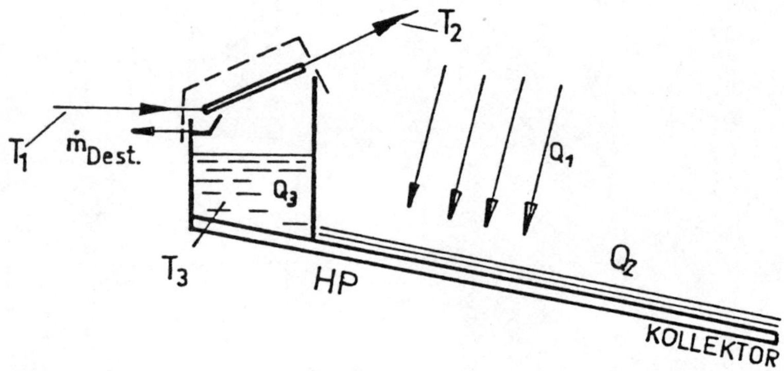

Q_1 = Insolation

Q_2 = Collector losses

Q_3 = Collected energy

T_1 = Temp. cooling water input

T_2 = Temp. cooling water output

T_3 = Temp. of the basin

Fig. 21 First Stage of the Desalination Unit

Fig. 22 View of the First Stage of the Desalination Process Showing some of the Heat Pipes

Fig. 23 Total View of the Second Stage of the Desalination Process

DORNIER
Solar Sea Water Desalination –
Scheme of Function of a Multistage Plant

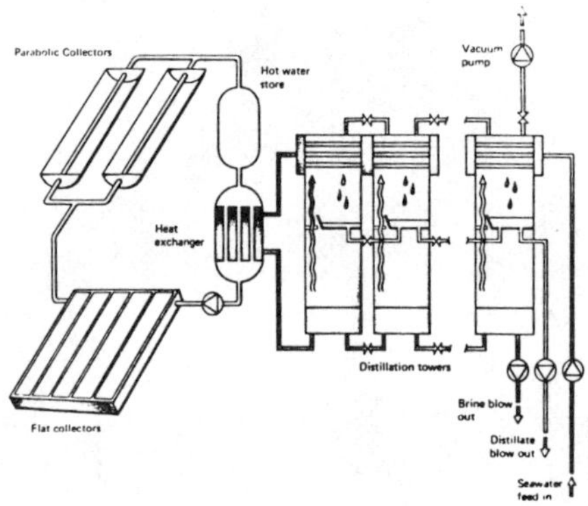

Fig. 24

Fig. 25 Total View of the Multistage Evaporating Plant

stage. the condensate is then sampled. In the second stage cold sea water(25°C) is pumped to a heat exchanger which is warmed up by the circulating air in the second stage; then this prewarmed water (45°) is fed to the cooler of the first-stage and is heated to about 60°C. This warmed water is used in the second stage to heat up the circulating air. Thus the humidity of the air increases and the additional water absorbed on the warm side is condensed on the cold side of the second stage.

This configuration has the following advantages:

- no corrosion problems within the collector area, for only the ends of the heat pipe are connected to the sea water and they are protected against corrosion.
  The basins are made of plastic.

- the system needs nearly no maintenance.

- The fabrication of the second stage is very simple.

Such a sea-water desalination unit has been mounted in Aquaba, Jordan and the following results have been obtained: with a daily insolation of 6, 5 kW/m^{2}day, about 5, 2 l/day m^2 have been produced. At a design point of 8, 5 kW/m^2 day about 9 l/m^2 day are expected. The total efficiency has been calculated, from measurements obtained in the field, to be about 60 %.

The system efficiency is defined as:

$$\eta = \frac{m_1}{m_2}$$

m_1 = measured amount of distilled water

m_2 = maximum ratio of distilled water if the total solar energy on the collector field installed can be converted for desalination (no thermal losses).

Fig. 22 shows the open basins, filled with sea-water and the ends of the heat-pipes inside the basins. In Fig. 23 the total view of a double stage-unit is shown (right side).

The second system, mainly developed by Dornier System is a multi-stage plant, where the seawater will be vaporised in 10 steps at different temperatures corresponding to different pressures inside a closed evacuated system (see Fig. 24). The input energy will be 60 kW. The expected output is calculated to be 10 m^3/day.Therefore the energy for 1 liter of distilled water will be 0, 14 kWh.

This facility is mounted in Mexico and will go in operation early in 1980.

Fig. 25 shows the total view of the multistage evaporating part, premounted and reintegrated in Germany.

SOLAR POWER GENERATION BY SOLAR PONDS

H. Tabor

The Scientific Research Foundation, Jerusalem, Israel

This note deals with the question of solar power and solar heat in large quantities. Specifically, it deals with solar ponds whose potential is highly exciting even though there are site constraints that limit the areas of applicability. However, in this respect, we should remember that, in the field of energy - as indeed in many other fields - there are not single solutions: every country - and indeed every locality - will provide its energy needs from a mix of sources and technologies, and every new and proven source or technology offered increases the options open to the decision-makers.

(Although this conference is entitled "Solar Energy in the 80's", history shows that a new technology may take from 15-25 years to become fully operational: thus we will probably have to wait for the nineties to feel the full impact of the technology discussed here).

December 16th, 1979 was declared, by the Israel Minister of Energy, a milestone in the history of harnessing the sun when he inaugurated a 150 kW thermal power station on the shores of the Dead Sea in Israel. This used a non-convecting solar pond as the heat source and a locally-built, low-temperature, organic-vapour Rankine turbine as the power plant. What was significant was that whilst both the pond and the turbine had been tested separately, this was the first time they had been "married" - with all the problems of interfacing successfully overcome.[1] Furthermore, whilst 150 kW is not large for a grid system, it represents a considerable jump in scale from previous solar units of a few kW, so that the next jump into the megawatt range can be approached with confidence. This 150 kW unit is the test step in the construction of a 5 MW plant now planned: the pond of 0.7 hectares area will be increased in two-steps to 1 sq. km. Indications are that 5 MW plants would cost about $3000 per kW installed, whilst larger plants would cost less, reaching an asymptotic value of approximately $2000 per kW installed for 40 MW or larger. These costs compare favorably with nuclear plants - when all the "hidden" costs of these plants are included - and with hydro-electric plants.

---

[1] The pond was constructed for another application in mind: it was "borrowed" for the present purpose. Though considerably smaller than would normally be used for 150 kW, it successfully supplied the heat needed to operate the 150 kW turbogenerator at full power.

The theory and technicalities of non-convecting solar ponds have been described many times and will not be repeated here. Suffice it to say that, like the OTEC (Ocean Temperature Energy Conversion) concept, absorption of solar energy takes place in a large body of water. In the non-convecting solar pond, this is artificially prepared and, by the imposition of a density gradient - by the dissolution of salt in increasing concentration with depth - water heated at the bottom of the pond remains there - and is insulated by the non-convecting water above. As a result, temperatures of over 100°C have been recorded at the bottom of such ponds, though 90°C is a practical working upper limit. The top of the pond is nearly at ambient temperature. This means that there is available a hot source and a cold sink - with a temperature difference in the region of 60°C - compared with 20°C, at the very best for an OTEC plant. Allowing for the temperature drops across the heat-exchangers, the net temperature difference for the solar pond plant is more than four times that for an OTEC plant, resulting in a similar ratio of thermodynamic Carnot efficiencies when used to produce mechanical or electrical power by means of a heat-engine, and very much smaller heat-exchangers.

The development of very high-reliability, low-temperature turbines now marketed by the ORMAT company - which has many thousands of such turbines working in over 40 countries and which has accumulated many millions of hours of operating experience - makes it practical to convert solar pond heat to power even though an even more economic use is for heating, where the Carnot limitation does not apply.

The solar pond approach attacks four of the major technical problems - apart from cost - pertaining to classical solar collectors:

(i) the low energy density of solar radiation calls for large collection areas: solar ponds can be made very large with negligible amounts of constructional materials.

(ii) the intermittancy of solar radiation. This calls for some form of energy storage if a continuous supply of energy is mandatory. The solar pond - because of its mass of water - has "built-in" storage. For the normal depths - of about 2 metres - the pond is insensitive to insolation variations over a week or more i.e. the output can follow the demand with little consideration of the supply: indeed the plant can be designed to provide peaks of power many times the nominal capacity. (Deeper ponds can be designed to provide seasonal storage i.e. from summer to winter).

(iii) the problem of dirt - that collects on the windows of classical collectors: a non-convecting solar pond has no windows. (Dirt entering the pond either sinks to the bottom - where it has negligible effect on performance or - if it remains suspended - may be removed by techniques developed for this purpose).

(iv) large solar installations raise the problem of transport of energy from a large collection area to a central zone of use. For example, in the "power-tower" concept, the energy is transmitted by reflected beam radiation: however, in most other thermal systems, a great deal of plumbing is needed. In the solar pond, transport of heat is particularly easy using the "decanting" method. It is known that, where a density gradient exists in a fluid, it is possible to move a layer horizontally without disturbing the layers above or below. It is thus possible to "decant" the hot zone of the pond, and transportation of this heat, over a distance of a kilometer or more, is feasible.

The cost figures given later show that a solar pond is a very low-cost source of calories at temperatures up to about 90°C: indeed they are so cheap that conversion to power, even with the low Carnot efficiencies, is feasible.

It is necessary to record that there are site-limitations to solar ponds i.e. they cannot be set up viably in all places. These limitations are:

(i) the pond, being horizontal, the sites should not be far from the equator i.e. they should be within about ± 40° latitude.

(ii) a source of cheap salt or brine must be available locally.

(iii) the site should be reasonably flat - to reduce the earth-moving work to a minimum.

(iv) unless complete sealing of the pond is assured, it would be inadvisable to construct a pond where there is underground fresh water near the surface.

(v) a cheap source of water (sea-water may be used) must be available to make up for evaporation losses.

(vi) it is not practical to make very small ponds, as the efficiency and economics improve with increasing size. The smallest practical size has not yet been established but is probably in the region of a hectare - for heat production - and several hectares for power production.

Where these site limitations do not apply, solar ponds appear to be an attractive method of harnessing solar energy on a large scale.

Efficiency of collection. Because half of the solar radiation is infra-red, which is absorbed in the first few cms of the pond i.e. does not penetrate to the bottom, - and allowing for other losses - the collection efficiency of a solar pond will - in practice - be of the order of 20%: this is lower than for other collectors but the unit area cost is an order of magnitude below that of flat-plate collectors.

Thermal yield. For a sunny area, such as Israel, with an annual insolation of 2000 kWH (thermal) per m^2 and a pond efficiency of 20%, the annual yield for a 1 km^2 pond is 400 million kWH - which is equivalent to burning 43,000 tons of fuel oil at 80% combustion efficiency.

Costs. The capital cost of a 1 km^2 pond is estimated to be US$13 million[2]: writing this off at 11.7% a year (8% interest and 15 years lifetime - which is conservative) and allowing for cost of make-up water ($200,000) and maintenance ($40,000) leads to a cost for heat delivered of $4400 per million kWH(t) ($1.29 per million BTU) i.e. competes with fuel oil at $41 per ton. This pond cost assumes the use of an expensive elastomer lining: the pond cost would drop about 40% for those sites where the lining could be omitted.

---

2 Salt and land are assumed to be free.

Electricity generation. We assume only 5% thermodynamic conversion efficiency (a figure already achieved) though it is expected that this will be raised to 8% in later units. Using the lower figure gives an annual yield of 20 GWH(e), equivalent to 5 MW installed at an annual load factor of 46%. (If the plant were to be used as a peaking plant - say 4 hours per day - the installed capacity could be raised to 13.7 MW).

Using very conservative figures of $1500/kW installed for the entire power-plant i.e. turbogenerators, heat exchangers and auxilliaries, leads to a capital cost - for a 5 MW unit, 1 km^2 pond - of $4100/kW installed and, allowing for the cost of make-up water and maintenance, the cost of power delivered at the bus-bars is 13.5 cents[3]. This is for a lined pond. If the lining is not needed, the installed cost falls to $3100/kW installed, and the cost per KWH falls to 10.5 cents. Power-plant costs are expected to fall and, if the thermodynamic efficiency can be raised to the expected 8%, bus-bar power costs will fall to as low as 5-6 cents - as shown in Table I.

The SOLMAT and ORMAT companies, that jointly designed and executed the 150 kW power station, estimate a final cost of about $3000/kW installed for 5 MW units, levelling off to $2000/kW for units of 40 MW or larger. Allowing for a slightly improved thermodynamic efficiency, the bus-bar power cost would then be of the order of 5 cents or less.

It is hoped that the first 5 MW plant will be "on line" within two years and 20 MW plants within a decade. It should be noted that, for a large central fuel-fired power station consuming, say 270 gms of oil per net KWH produced, the fuel component of power cost is 0.2C cents where C is the cost of a barrel of fuel oil delivered. Thus with oil costs - at the time of writing - of $30-40 per barrel, the fuel component cost is 6-8 cents: this does not include the cost of the generating plant. For small local diesel-generators, power costs frequently range between 15 and 30 cents/KWH, primarily because of the high maintenance and operating costs of these plants.

Thus solar pond power may compete with central power production using fuel oil at current and expected fuel prices: it seems highly competitive with local diesel generators.

A rather unexpected bonus, for developing countries, is that the cost of solar pond power stations - which decreases with size for small sizes - reaches an asymptote at around 20-40 MW. This means that small countries like Israel can increase their installed capacity in small steps as the demand develops, rather than in the large steps - of 600 MW or more - that is recommended for nuclear or fossil-fueled stations.

Conclusions. For those areas where solar ponds are practical, this represents an extremely interesting possibility as an energy source in place of oil or other fossil fuels. For thermal applications - and a large amount of primary energy in many countries goes for low-temperature thermal processes - the solar calories provided by solar ponds appear lower in cost than calories from almost any other source. District heating is one application, provided the limitations of the site, including availability of cheap land[4], are taken into account. Central air-conditioning in warm climates is another

---

[3] Using 11.7% annual charges on capital as before.

[4] A new development under study is the construction of ponds that "float" in the sea - or in lakes. This could avoid the use of land where this is expensive.

possibility. Desalination - using low-temperature multi-effect distillation units as developed in Israel is another possibility[5]. An earlier paper estimated a pond of 0.3 km^2 for a plant yielding 4000 m^3 per day (10^6 gpd) average throughout the year in a sunny climate.

The areas of the world where solar ponds are practical (and where there is a need for thermal or electrical energy) have not been fully delineated but it is clear that the overall potential is very great: whilst solar ponds may not contribute too much in the highly industrialised countries - where the site conditions are not favorable - the contribution of solar ponds to developing countries short of indigenous fossil fuels could be of considerable significance.

---

5 A demonstration desalination unit has been set up adjacent to the 150 kW power station.

TABLE 1 Power Costs From Solar Ponds

(Solar input: 2000 KWH/m^2 yr. 1 km^2 pond area)

	LINED POND US$13/$m^2$		UNLINED POND US$8/$m^2$	
Power-plant(1) cost $/kw installed	1,500	1,000	1,500	1,000
ANNUAL COSTS (Based on annual charges on capital of 11.7% i.e. 8% interest and 15 years lifetime)				
Pond	1,521,000	1,521,000	913,000	913,000
Water	200,000	200,000	200,000	200,000
Maintenance	100,000	100,000	100,000	100,000
Power-plant	877,500	585,000	877,500	585,000
Total $	2,698,500	2,406,000	2,090,000	1,798,000
BUS-BAR POWER COSTS(2)				
At 5% thermodynamic conversion: annual yield = 20 GWH(3)				
Cents/KWH	13.5	12.0	10.5	9.0
At 8% thermodynamic conversion: annual yield = 32 GWH(3)				
Cents/KWH(4)	8.1	7.2	5.9	5.3

(1) Plant refers to turbogenerators, heat exchangers and all auxilliaries.

(2) Excludes station operating costs

(3) Assumes pond collection efficiency = 20%

(4) Includes allowance for reduced size heat exchangers

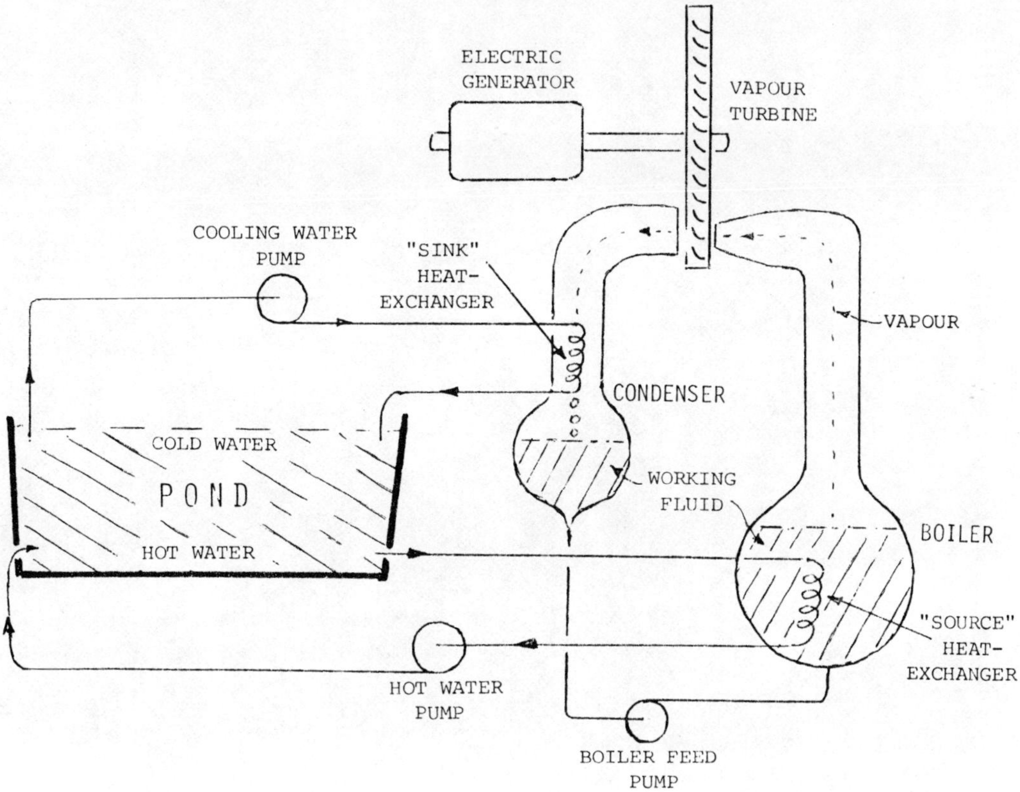

Schematic Diagram of Solar-Pond Power Station

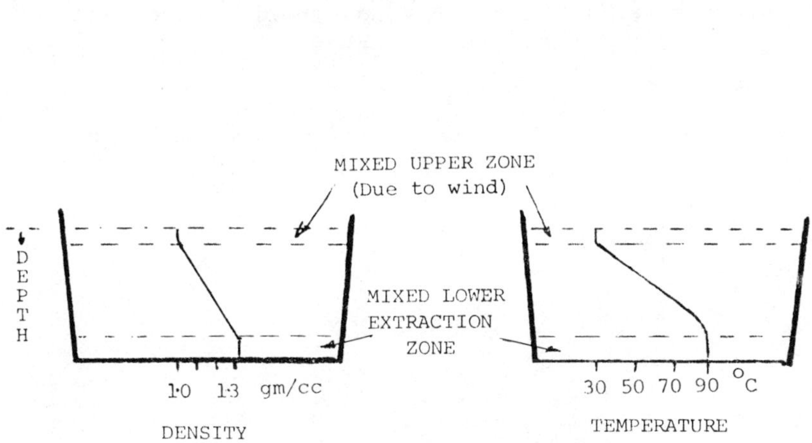

Density and Temperature Gradients in Solar Pond

General View of the 150kW Solar Pond, Thermal Power Station on the Shores of the Dead Sea in Israel

PHOTOVOLTAIC CONVERSION OF SOLAR ENERGY IN THE 80s

Y. Chevalier

COMES, 208, rue Raymond Losserand, Paris 14, France

ABSTRACT

Among the many possible ways of converting solar energy into useful energy for mankind, photovoltaic conversion appears a highly desirable way for producing electricity for small and medium scale applications, both in industrialized and developing countries :

- Photovoltaic devices, known as solar cells, convert directly sunlight into electricity with no intermediary processes.

- Solar cells are simple : they involve no moving parts and little if any maintenance.

- Photovoltaic systems are capable of operating efficiently in a variety of applications ranging from small, low power devices such as remote instruments, to dispersed systems (residences, schools, hospitals, pumping stations) to grid-connected central power plants.

- Costs have dropped steadily by 20 % in the past five years and are expected to follow this trend in the next years.

- More than 2 peak -MW have been already installed for powering of numerous and extremely different applications all around the world. When the cost decreases, photovoltaic conversion will become cost effective for a greater number of applications.

KEYWORDS

Photovoltaic cells, modules systems, applications.

PHOTOVOLTAIC CELLS AND MODULE TECHNOLOGIES

Present state of the art :

Most commercial photovoltaic modules available nowadays are based on single-crystal silicon technology.

As shown in fig. I, highly pure polysilicon is first obtained from silicon dioxyde. This electronic-grade silicon is then pulled by Czochralski method in order to form a single-crystal ingot. I/2 mm thick wafers are saw-cut from the ingot and processed (junction formation, metallization, anti-reflective coating) ; the formed solar-cells are capable of delivering 0.5 V (almost independant on the illumination) and about 30 mA/cm2 under Air Mass I illumination (I kW/m2).

Finally, in order to protect the cells against environment and to obtain output voltages and powers compatible with the commercial applications to be powered, the cells are interconnected and encapsulated.

The final product, known as solar module, is a device ready to be installed and capable of delivering powers ranging from 2 W to 40 W.

Short and medium-term technologies

As described previously, the present technology is using both expensive base material and batch process cell fabrication methods. Production costs can substantially be lowered for all stages of production :

- Cheap "solar grade" silicon is under development.
- Good quality poly or semi-crystalline silicon, either obtained by ribbon, casting, chemical vapor deposition technics, is now available.
- Automation of the processes involved in the cells and modules fabrication will contribute to decrease present costs. Besides the reduction of production costs, the cost/peak-W of photovoltaic devices can be still reduced as conversion efficiency will improve : I8 % efficiency cells have already been obtained in laboratory.

Long term technologies

Single-crystal and/or polysilicon cells need to be at least 40 microns thick in order to efficiently convert solar energy. Thin cells which have been shown to be technically feasible consume much less base material.

Cells made from compound materials (Cd S, Ga As, etc.) or amorphous silicon will certainly be the I990's cells if they could present efficiencies comparable to those obtained with mono or polysilicon and if degradation problems could be solved.

Finally, mass production (more than IOO MW expected in I985) will also greatly contribute to the reduction of the cost per peak-watt.

SYSTEMS AND APPLICATIONS

Since the fabrication of the first silicon solar cell in I955, thousands of applications have been powered by photovoltaic systems.

In the past years, as the cost/peak-watt was still rather high, cost-effective photovoltaic installations were restricted to isolated applications requiring low-power (in the I W to I kW range). For such applications, the cost of replacement of batteries, transportation of fuel, or remote power lines makes conventional supply prohibitive.

The first photovoltaic applications include remote educational TV-sets, weather monitoring stations, navigational lighting, radio-beacons ; radio, television, microwave booster and repeater stations.

It has to be noticed that, for most of these applications, solar electricity must be stored in batteries in order to be available for continuous or not sun-dependant operation.

Since the cost of photovoltaic systems drops, larger applications are to be powered by solar electricity. Most of them are particulary suitable for rural areas of developing countries.

Developing countries applications

The increased emphasis on rural development in developing countries will mean increased demand for decentralized power sources, since most rural areas cannot be reached soon by an electric grid.

In the frame of conventional electrification schemes, priority is given to the cities and the installation of an electric grid for scattered populations in vast land areas is very costly.

Massive introduction of solar cells in the sunny areas of the developing countries would include important social benefits as the basic needs of the rural populations (education, communication, health, agriculture and drinking water supply) can be satisfied by photovoltaic systems :

- 20 W are sufficient to power a TV-receiver which makes possible to improve the quality of education of the children in a village and to improve illiteracy. Thousands of such TV-receivers are to be installed in the next years ;

- A rural telephone, requiring approximately the same power, allows a whole community to get out of its isolation ;

- A IOO liter refrigerator for rural health dispensaries only requires a I50 W installation ;

For these small applications, cost of photovoltaics does not even play a significant role as it is hard to compute the social benefit obtained.

On the other hand, as water pumping is generally provided by Diesel engines, photovoltaic generators have to compete with such power supplies. Taking into account both the high maintenance and fuel supply costs of traditional generators and the fact that in most cases electrical storage is not necessary for pumping, photovoltaic pumps are currently cost-competitive for applications up to IO kW in remote regions.

Photovoltaic technology could also play an important role in slowing down rural exodus, providing the basic energy needs to scattered populations. In this scheme the modular nature of photovoltaic systems permits the users to gain experience with a relatively small initial investment. As the local community grows, system elements could be designed to allow the development of local "grids". Loads can grow with supply, meaning essentially full amortization of the investments.

As an example, the French Commissariat à l'Energie Solaire, recently installed a IO kw photovoltaic generator, in San, Mali for powering a rural hospital.

Such medium-size installations should become very popular in the next years since photovoltaic kWh cost is already of the order of magnitude of national electricity companies kWh cost in some developing countries.

Industrialized countries applications

In the short and medium terms, since industrialized countries are provided with electricity grids reaching almost all communities and since they are not located in the sunniest parts of the world, photovoltaic generators will only be cost competitive for specific applications such as telecommunication, cathodic protection, control of corrosion, railroad and highway signals, telemetry and small water-pumping.

In some rare cases, isolated farms and residences could be already economically equipped with photovoltaic generators.

As the present cost ($ I0/peak-watt) of photovoltaic modules is expected to decrease by one order of magnitude in this decade, numerous residential and commercial applications will become cost-effective : neglecting the electricity grid kWh cost and only considering the investment costs ($ 2/peak-watt for photovoltaics and $ 30 000/ mile for grid connection), buildings located at more than one mile of the grid could be economically equipped with a I5 kw photovoltaic generator, delivering 30 000 kWh per year in Southern Europe.

Nevertheless, it should be mentioned that, although this market is important, photovoltaic electricity is not expected to play a significant role in industrialized countries energy consumption in this century.

CONCLUSIONS

"Photovoltaics people" are now entirely confident in the predicted market and cost goals : installed peak-watt should not cost more than 2 dollars by the end of the decade while world market should be larger than I00 MW in I985. In the past two years, public funding of photovoltaic research and development has substantially increased all over the world : in I979, I25 million dollars have been spent by United States, 30 millions by Europe (a third of it beeing spent by France).

On the other hand, industrial investments are increasing dramatically : as an example, Arco-solar just decided to invest 25 million dollars on amorphous silicon and I2 million on cast silicon research and development.

In France, large companies (CGE, THOMSON-CSF, RTC, LEP, LEROY-SOMMER, ELF, CFP) are already involved in the manufacturing of photovoltaic products.

Still a lot has to be done, but one can be certain already that future generations will consider photovoltaic conversion of solar energy as a "classical" method for generating electricity.

REFERENCES

- Durand, H. (I979)."Photovoltaic Solar Energy Conversion". Conference at the Royal Society of London

- Palz, W. (I979). "Photovoltaic for development" Proc. Photovoltaic Solar Energy Conference, Berlin.

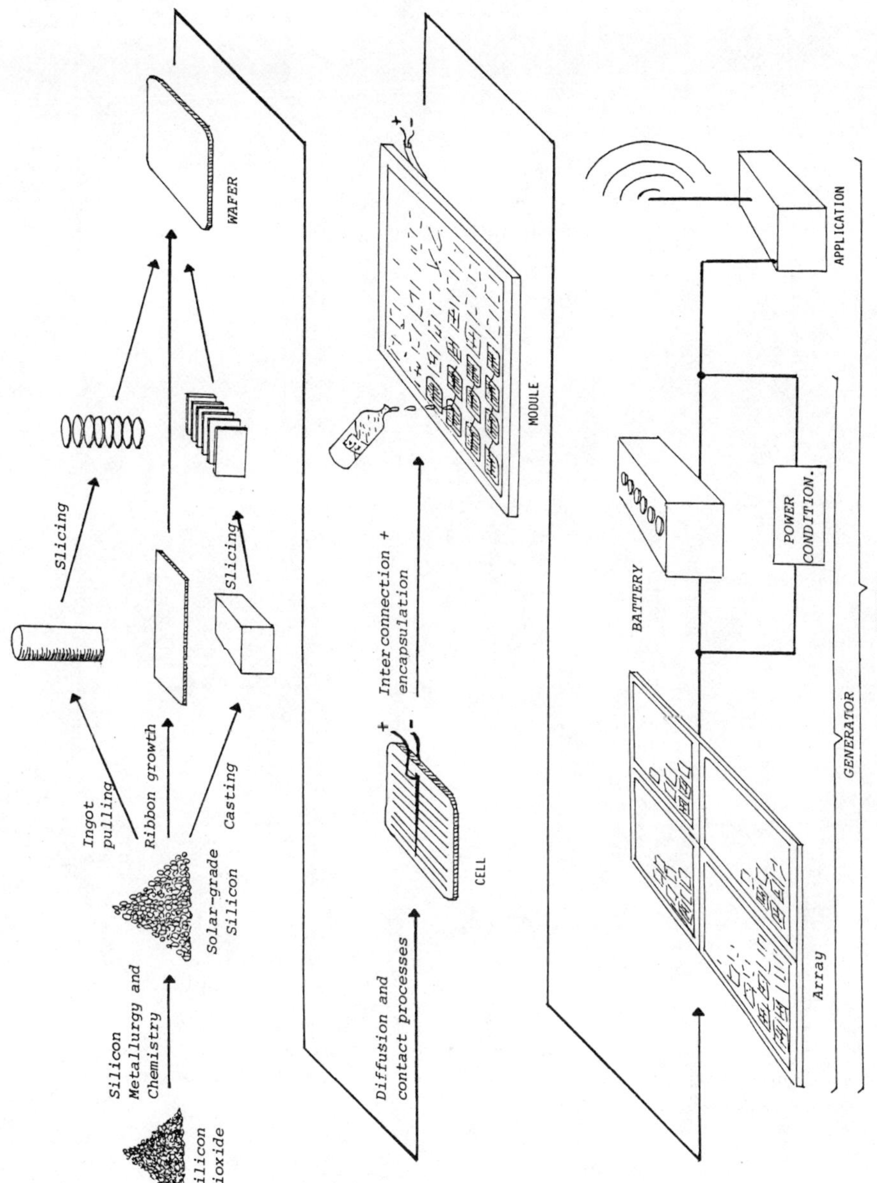
Silicon dioxide
Silicon Metallurgy and Chemistry
Solar-grade Silicon
Ingot pulling
Ribbon growth
Casting
Slicing
Slicing
WAFER
Diffusion and contact processes
CELL
Interconnection + encapsulation
MODULE
Array
GENERATOR
BATTERY
POWER CONDITION.
APPLICATION
SYSTEM

Fig. 1

THE DEPARTMENT OF ENERGY'S SOLAR ENERGY RESEARCH AND DEVELOPMENT PROGRAMME

E. G. Bevan

Energy Technology Division, Department of Energy, Thames House South, Millbank, London SE1P 4QJ, UK

ABSTRACT

The background to the Department of Energy's solar energy R and D programme is sketched, with particular reference to the ETSU (Energy Technology Support Unit) assessment published as Energy Paper No 16, and to the on-going work of other Government Departments. The aims of the Department's 4-year solar energy R and D programme, announced in February 1977, are summarised and the approaches adopted in seeking suitable proposals from industry, research organisations, universities and polytechnics are described. Reference is also made to the Department's international activities through the EEC and the IEA. Progress to date is reviewed in the various applications covered in the programme, viz solar water heating, active solar space heating, passive solar spacc heating. Finally the possible future contribution to UK energy supplies from solar energy is considered, with particular reference to the Department's energy projections and to some recent scenario analyses carried out by ETSU.

KEYWORDS

Department of Energy UK; solar heating R and D; funding arrangements; programme aims; programme content.

BACKGROUND

1. A number of Government Departments engage in energy research and development; those involved with solar energy R and D in particular include the Department of Energy, the Department of Industry, the Department of the Environment, the Overseas Development Administration, the Ministry of Agriculture, Fisheries and Food, and the Research Councils, mainly the Science Research Council.

2. The Department of Energy holds the central position in Government for all energy R and D, and in addition to its responsibilities for conventional fuels and nuclear energy it is responsible for:

- R and D into novel sources of energy;
- co-ordination of energy R and D within Central Government.

The Department is advised on its alternative energy R and D programmes by the Advisory Council on Research and Development for Fuel and Power (ACORD), chaired by the Department's Chief Scientist, Sir Hermann Bondhi, except for work relating to tidal power from the Severn Estuary where advice is given to the Department by the separately constituted Severn Barrage Committee. The Department's R and D assessment and project management capability is strengthened by the Energy Technology Support Unit (ETSU) set up at Harwell in 1974. Through ETSU the Department of Energy has now established substantial R and D programmes on wind, wave and geothermal energy, on solar heating, biofuels and tidal power.

3. One of the first studies commissioned from ETSU by the Department was an assessment of the potential contribution to UK energy supplies from solar energy; this assessment covered a wide field including solar heating, solar power and biomass. ETSU was asked also to make recommendations on an appropriate research and development programme to be funded by the Department of Energy, in the light of work then being supported in this field by other bodies. The results of the ETSU assessment were published in 1976 as Energy Paper No 16, "Solar energy: its potential contribution within the United Kingdom", available from HMSO. The paper's main conclusions may be summarised as follows:

(a) Solar energy (in the form of low grade heat and fuels from biomass) could make a contribution totalling some 2 per cent of current energy demand within 25 years, with minimal disturbance to existing lifestyles and institutions.

(b) In the longer term, with vigorous exploitation, this contribution might be increased by a factor of 10.

(c) Exploitation on any significant scale will be slow, and limited by the difficulties of market penetration.

(d) Because of the UK's climatic conditions power generation by photovoltaic conversion or focussing collectors is unlikely to be economic in the foreseeable future.

4. The conclusions and recommendations of the ETSU assessment were considered by the Department of Energy against the background of on-going R and D programmes funded by other Government Departments and the Research Councils. Following advice from ACORD the Department announced in February 1977 a 4-year solar energy R and D programme costing £3.6 million, bringing the total Government funding for solar energy R and D over that period to £6 million. The £3.6 million was allocated as follows:

- £2.8 million on solar water and space heating
- £0.2 million on solar data
- £0.6 million on solar biological systems

In February 1979 a further £1.5 million was allocated to the solar biological programme on the basis of encouraging results from desk studies. Work on solar energy continues to be supported by other Government Departments; this includes complementary work on solar heating by the Building Research Establishment of the Department of the Environment, R and D on photovoltaics supported by the Department of Industry (mainly to assist industry with exports) and research on photochemistry and photobiology supported by the Science Research Council and the Agricultural Research Council (mainly in universities).

PROGRAMME AIMS AND CONTENT

5. This paper will concentrate on the Department of Energy's solar heating R and D programme, only brief reference will be made to the biomass work. The solar heating programme has 3 main aims. These are:

- to encourage firms to optimise solar water heating systems, to define their performance and to demonstrate their value in real life situations;
- to determine the feasibility and long term costs of a selection of solar space heating systems by design studies, laboratory trials (including test houses) and demonstrations;
- to encourage firms to develop individual components for solar heating systems against system and cost criteria, and to estimate the long term costs of mass production and installation of these components.

The programme has been built up through an appropriate balance of projects placed with industrial firms, contract research organisations, and universities and polytechnics. The general strategy adopted has been to support industrial initiatives on a 50:50 funding basis and to fund, up to 100 per cent, other relevant research and development projects and monitoring exercises necessary to realise the programme aims.

6. To date we have 50 projects in the programme. These fall under 3 broad headings as follows:

(a)	WATER HEATING SYSTEM & COMPONENT DEVELOPMENTS (INC. SOLAR DATA)	29 projects
	Total Dept. of Energy commitment	£1.3 million
	Total commitment from others (industry, local auth., EEC etc)	£0.5 million
(b)	ACTIVE SPACE HEATING	14 projects
	Total Dept. of Energy commitment	£1.0 million
	Total commitment from others	£0.6 million
(c)	PASSIVE SPACE HEATING	7 projects
	Total Dept. of Energy commitment	£0.4 million
	Total commitment from others	£0.6 million

The programme total commitment to date therefore is £4.4 million of which £2.7 million is contributed by the Department and £1.7 million by the other participants. A summary of the details of the 50 projects is contained in Annex A. This annex gives a brief guide to the spectrum of work being undertaken, the nature of the projects, the spread of contractors and the number of projects where joint funding has been arranged.

7. There are a number of points to which attention should be drawn. First, monitoring the performance of solar heating systems in dwellings, schools, hospitals and other residential buildings forms an important part of the programme and such projects are commissioned as opportunities arise. Generally

speaking the Department funds the total cost of monitoring the installation but in appropriate cases may also make a contribution towards the capital cost of the system itself. One of the prime functions of the ETSU solar energy management team is to ensure that comparable data is obtained from all these monitoring exercises and that the analysis is carried out in a uniform manner so that the programme results will be of maximum use - both in a national and international context. Likewise with system modelling projects ETSU are endeavouring to ensure that, where possible, common methods and approaches are employed. Secondly, to aid establishing both quality and consistency in work in the monitoring, modelling and performance testing areas we have built upon the knowledge and experience of certain organisations recognised for their expertise. Among these are the Building Research Establishment (BRE), the Building Services Research and Information Association (BSRIA), the Solar Energy Unit of University College Cardiff, Faber Computer Operations, and the Architecture Department of Strathclyde University. Thirdly, unlike the case with other renewable energy sources an industrial base does exist in the solar field, though at present it is comparatively small. Several industrial companies with sizable research and development facilities and well-organised marketing operations were already evaluating solar heating and considering its market potential when the Department launched its R and D programme. It was with such companies that we commenced building up the industrially based part of our R and D programme. To date the programme includes 19 projects in industry, 15 on a shared cost basis and 4 totally funded by the Department. These figures illustrate the considerable success in forming collaborative links with manufacturing industry.

8. International collaboration plays an important part in the Department's programme. It enables a broader data base to be achieved, more generally applicable reference designs to be devised, and it offers opportunities for cost and task sharing. The programme includes collaborative projects through the EEC and the IEA. Our Prototype Test Facility (PTF) located at, and supplied by, BSRIA is an integral part of the EEC programme in the solar heating area. Similar test facilities are being built in other EEC countries; it is worth reporting that Britain's PTF was the first of these to be completed and it is now one of the centre pieces of our solar heating programme. The new 4-year solar R and D programme of the European Commission totals 46 million units of account, 8.3 million of which has been allocated to solar heating. There is clearly scope for increased UK participation in the EEC programme and the Department encourages contractors, where appropriate, to seek funding from the European Commission. The Department's involvement in the solar R and D activities of the IEA is detailed in Annex B; here advantage is being taken of the opportunities for task sharing, particularly in performance testing aspects.

PROGRESS TO DATE

9. First it must be emphasised that this is a 4-year programme and that it still has a full year to run. Although a large proportion of the funds originally allocated to solar heating R and D has now been committed, work on many of the projects is only in the initial stages. A high proportion of the projects span 2 or 3 years in duration and therefore it will be some time before the bulk of the results can be analysed. It is not possible, therefore, to give any major conclusions in this paper. However, we can say that the uncertainties in previous assessments relating to solar heating in the UK have been considerably reduced. This process will continue as further results come forward and the work currently in hand should form a sound base for future developments. It is being confirmed that reliability is of paramount importance if market penetration is to occur on the scale required to make a significant contribution to energy supplies. The

work being carried out at University College, Cardiff, on collector testing and that in support of the work for the BSI is important in this context. It is also becoming clear that much more emphasis must be placed in future on system development; the PTF will have a major role to play in this area.

Water Heating

10. Our studies to date indicate that in general, at rates of production justified by the current market, solar water heating is too expensive by a factor of 2 or 3 to make investment economic against current fuel prices. A number of technical and engineering areas have emerged from the work to date, each of which promises improvements in cost and performance in the range 5 to 30 per cent. Taken together these evolutionary developments could bring down the total system costs substantially. Apart from increased emphasis on total system development, further R and D needs to be done on high performance collectors. We are, therefore, supporting new work in collectors of the evacuated and semi-focussing types.

Active Space Heating

11. Active solar space heating is more complex than solar water heating and its development is on a much longer time scale. Our current field trials and design studies are beginning to indicate which specific lines of development should be undertaken in future projects. For instance air systems might provide an attractive solution to some of the current problems; such systems could be particularly compatible with low grade heat collection. Heat storage, of course, is extremely important in active systems and, as other speakers at this conference have pointed out, some solution must be found. ETSU are now engaged in assessment studies in this area and the programme already includes a storage element.

Passive Space Heating

12. Passive systems could be the first form of solar space heating to be economic in the UK. More emphasis will be placed on this area in future. It is likely that passive systems based on conventional house designs will become cost-effective in advance of active systems for dwellings built at densities of 30 houses per hectare or less. Unfortunately few opportunities exist in the passive area for R and D projects since such systems are closely tied to building design. Milton Keynes, however, is providing us with opportunities to monitor a number of passive heating concepts.

THE CONTRIBUTION FROM SOLAR ENERGY IN THE UK

13. The ultimate contribution to UK energy supplies might be:

Solar Water Heating	10-20 mtce per annum
Solar Space Heating	40-80 mtce per annum.

But such figures might never be realised - if they are, it probably will not be until well into the next century. This conference is entitled "Solar Energy in the 80s", in the UK the contribution to energy supplies this decade from solar energy is likely to be negligible and indeed it is likely to remain small till the end of this century. The Department's latest energy projections to the year 2000, which up-date those given in the 1978 Energy Policy Green Paper, do not include any contribution from renewable sources, other than hydro. This is because, on present evidence, they seem unlikely to contribute significantly to supply or be economically preferable to other fuels over the period to the end of this century.

Next century, however, solar energy could be making an increasing and valuable contribution to heating requirements.

14. Recently ETSU have carried out an appraisal of energy technologies for the UK, based on an exploratory scenario for the next 50 years. The long term assumptions in this scenario are broadly similar to those used in the Department's energy projections. This ETSU work has been published as Energy Paper No. 39, "Energy technologies for the United Kingdom - An appraisal for R, D and D planning", available from HMSO. This appraisal suggests that the upper limit of the contribution from solar water heating by the year 2000 is 4 mtce per annum, but in practice it is likely to be something less than 0.5 mtce per annum. On the ETSU scenario the contribution from solar space heating is not likely to exceed 1 mtce per annum until after the end of this century. If the cost of solar water heating systems could be brought down by a factor of 2, say £400 for an average 3-bedroomed dwelling, with average hot water consumption, then solar water heating could compete with gas at 40p/therm or off-peak electricity at 60p/therm. The ETSU exploratory scenario suggests, however, that gas and electricity prices will not reach these levels until perhaps the second decade of the next century.

15. As far as the contribution from solar biomass is concerned the ETSU assessments have indicated that the ultimate technically feasible contribution to UK energy supplies in the form of biofuels is around 60 mtce per annum of raw material (wastes and crops). It is unlikely, on an orthodox view of the future, that energy from crops will make any significant contribution before 2025, but fuels from wastes could contribute up to 4 mtce per annum by the year 2000 and perhaps up to 7 mtce per annum by 2025.

ANNEX A: DEPARTMENT OF ENERGY SOLAR HEATING R AND D PROGRAMME

TABLE 1: Solar Heating System and Component Development Projects

Object	Project Type	Contractor	Funding
Optimised reference design for domestic water heating system	System Dev.	Consultant	100% D/En
Study of installation problems and component interactions in water heating systems	System Dev. & Field Trial	Industry	100% D/En
Monitoring of water heating systems in 13 local authority houses in Southwark	Field Trial	Local Authority	100% D/En
Monitoring and model validation of water heating systems in 2 housing schemes in Birmingham	Field Trial	Local Authority	Shared
Monitoring of large water heating system for hospital complex, Torbay	Field Trial	Local Authority	Shared
Monitoring and evaluation of water heating systems in 40 houses in Basingstoke, 20 in Lincoln and 15 in York	Field Trial	Research Assoc.	100% D/En
Development and evaluation of superplastic aluminium collectors	Component Dev. & Field Trial	Industry	Shared
Development and evaluation of components and system for large scale water heating in a hospital	Component Dev. & Field Trial	Industry	Shared
Development of fixed semi-focussing collectors	Component Dev.	Industry	Shared
Design and development of low life collector panel for domestic and industrial water heating	Component Dev.	Industry	Shared
Development and manufacture of cost-effective collectors of conventional design	Component Dev.	Industry	Shared

TABLE 1 (Cont'd)

Object	Project Type	Contractor	Funding
Development and testing of collectors for water heating and computer simulation of thermosiphon circulation	Component and System Dev.	Industry	Shared
Investigation of manufacturing techniques for collectors and their heat transfer characteristics in domestic water heating systems	Component Dev.	Industry	Shared
Investigation of tubular solar energy collectors and their manufacture	Component Dev.	University	100% D/En
Determination of durability and performance of selected surfaces in flat plate collectors	Component Dev.	Industry	Shared
Assessment of wind-loadings and fixings for roof-mounted collectors and fabrication of preferred concepts	Component Dev.	Industry	100% D/En
Development of integral pump/ controller for solar heating systems	Component Dev.	Industry	Shared
Performance testing of solar collectors and development of standard test procedures	Component testing	University	100% D/En (IEA task)
Development of collector testing procedure for transient conditions	Component testing	University	100% D/En
Provision of solar equipment testing service	Component testing	University	100% D/En
Development and manufacture of honeycomb collector	Component Dev.	Industry	Shared
Study of potential for solar water heating in non-domestic process applications	Info.	Consultant	100% D/En

TABLE 1 (cont'd)

Object	Project Type	Contractor	Funding
Application of non-imaging solar concentrator technology to UK systems	Modelling	Research Est.	100% D/En
Optimisation of solar energy systems by multivariate analysis	Modelling	University	Shared (EEC)
Construction of theoretical models for water heating in students flats, Cardiff	Modelling	University	100%
Assessment of economic influence on the potential for domestic solar water heating	Info.	Consultant	100% D/En
Theoretical analysis of thermal performance of water heating systems and assessment of potential in the textile industry	Info.	Polytechnic	100% D/En

TABLE 2: Solar Data Projects

Object	Project Type	Contractor	Funding
Preparation of Insulation Handbook and application of existing data for solar energy purposes	Solar data	Met. Office	100% D/En (IEA task)
Measurement of solar radiation on inclined surfaces and to development of new instrumentation	Solar data	University	100% D/En

TABLE 3: Active Solar Space Heating Projects

Object	Project Type	Contractor	Funding
Prototype domestic space and water heating system	System Dev. and Field Trial	Industry	Shared
Prototype domestic air space heating system	System Dev. and Field Trial	Industry	Shared
Testing of different space heating systems in 9 houses at Milton Keynes	System Dev. and Field Trial	Industry	Shared
Monitoring performance of large space heating system in home for elderly persons, Cheshire	Field Trial	Local Authority	Shared
Monitoring performance of space and water heating system at Port Isaac school	Field Trial	University	100% D/En
Construction and operation of Pilot Test Facility (PTF) for solar heating systems	System Dev.	Research Assoc.	Shared (EEC)
Design of small solar-heated industrial building	Design Study	Consultants	Shared
Study of combined solar and heat pump domestic heating systems	Design Study	University	100% D/En
Design of Milton Keynes Mk II domestic space heating system	Design Study	Polytechnic	100% D/En
Design of evacuated collector space heating system	Design Study	Industry/ Research Assoc.	100% D/En (IEA task)
Construction and study of basic types of air collectors for space heating	Component Design and Modelling	University	100% D/En
Design, manufacture and test of phase-change thermal storage unit	Component Design and Modelling	University	Shared (EEC)
Theoretical investigation of long-term storage in the ground	Modelling	University	100% D/En

TABLE 3 (Cont'd)

Object	Project Type	Contractor	Funding
Theoretical investigation of performance of solar heating/ cooling systems using evacuated collectors	Modelling	Industry	100% D/En (IEA task)

TABLE 4: Passive Solar Heating Projects

Object	Project Type	Contractor	Funding
Monitoring of passive solar gain in 60 houses at Pennylands, Milton Keynes	Field Trial	Local Authority	Shared
Monitoring of 8 direct gain passive solar houses at Linford, Milton Keynes	Field Trial	Local Authority/ University	Shared
Monitoring of 5 houses at Aylesbury, 4 designed with various features for passive solar gain	Field Trial	Industry	Shared
Monitoring of 14 low energy houses at Bebington, Cheshire, designed for passive solar gain	Field Trial	Industry	Shared
Monitoring of experimental solar heated dwelling	Field Trial	University	100% D/En
Computer simulation of passive solar systems	Modelling	University	100% D/En
Review of US passive solar heated work	Info.	Consultant	100% D/En

ANNEX B: DEPARTMENT OF ENERGY'S PARTICIPATION IN IEA SOLAR HEATING PROJECTS

The Department of Energy is funding 100 per cent of the cost of UK participation in the following IEA solar heating tasks. The Department's support of these projects to date totals just over £0.25 million.

Task	Participant
Task I: Investigation of System Performance	Industry
Task II: Performance Testing of Solar Collectors	University
Task III: Development of Insulation Handbook	Meteorological Office
Task IV: Use of Meteorological Data for Solar Energy	Meteorological Office
Task V: Investigation of performance of solar heating systems using evacuated collectors	Industry/Research Assoc.

US INTERNATIONAL COMMERCIALIZATION PROGRAMS FOR SOLAR ENERGY

R. C. Spongberg

International Programs Division, Solar Energy Research Institute, Golden, Colorado, USA

ABSTRACT

Paper discusses the international commercialization activities of the U.S. Department of Energy and the Solar Energy Research Institute. The programs underway deal with: 1) assessment of the market potential for commercially ready products, technologies and services; 2) acquisition, storing and dissemination of information which will facilitate commercialization; and 3) promotion of solar capabilities. It concludes that cooperative efforts now will provide major benefits to the international solar industry sooner.

KEYWORDS

International commercialization; bilateral programs; cooperative research.

INTRODUCTION

On October 26, 1974 with the enactment of the Solar Energy Development and Demonstration Act of 1974, the United States took its first major steps into the Solar Age. Since then, the U.S. has expended many dollars in research, development and demonstration in its effort to assure the utilization of solar energy as a viable energy resource. There has been some criticism of the direction of the U.S. Solar program from both inside and outside of the U.S. This has been in the form as such cliches as "The U.S. spends money on Solar energy like they think they can produce a child in one month by getting nine women pregnant" or "Solar will become a viable energy resource when the energy derived directly from the sun is equivalent to the energy derived from that being generated by burning all of the reports written about solar energy in the U.S." In spite of the criticisms, however, it is apparent that the multidirectional effort in the United States has and is helping to produce an accelerated interest in solar energy outside of the U.S. It has also allowed other nations to rifle in on programs with specific national benefits. It has also provided the world with a significant information base that is being aggresively disseminated and utilized.

This is in keeping with the above mentioned directives from Congress relative to the international arena. The Act specifically requires that there will be cooperation and participation jointly with other nations in such activities as:

- international, bilateral, or multilateral research projects in the field of solar energy.
- agreements and programs which will facilitate the exchange of information and data.
- the development and maintenance of programs of education.

INTERNATIONAL ACTIVITIES

Armed with the mandate, the Department of Energy (DOE) and its Solar Energy Research Institute (SERI) have, together with other nations, undertaken a number of projects to hopefully reduce the time when solar energy will be considered as commercially viable as nuclear and the other, non-renewable, energy resources. The international activities of D.O.E. can be grouped into three areas. They are:

1. Cooperative research and development with other industrailized or capital surplus developing countries.
2. R and D in energy systems for the developing countries.
3. Commercialization of appropriate and technically ready renewable energy systems.

In the first area, the following cooperative efforts have already been initiated.

- Multilateral cooperation with the International Energy Agency and with the North Atlantic Treaty Alliance Committee on the Challanges to the Modern Society, and
- Bilateral cooperation with Saudi Arbia, Spain, the U.S.S.R., and France.

Future cooperative arrangements are being discussed with Australia, Israel, Japan, Mexico, Brazil, Italy and Korea.

There is clearly a need for R and D to develop energy technologies specifically designed to operate in the climatic and working environments of developing countries. Such programs are being assessed and initiated. In September 1977, D.O.E. initiated the planning and management of programs of cooperative analysis of energy needs, resources and uses with Egypt and Peru.

In each of the above mentioned areas, SERI is heavily involved in providing the necessary technical and management support.

With regard to commercialization, it is quite apparent that certain solar technologies will probably be economically viable in developing countries earlier than in the U.S. This is primarily due to the artifically depressed commercial fuel prices; it is also due to infrastructional and institutional barriers which have to be overcome before solar can become an attractive alternative. Thus, the development of markets outside of the U.S. may be one of the necessary first steps to stimulating the use of solar within the U.S. With this in mind, the D.O.E. has been working with other U.S. government agencies, with SERI, with U.S. industry associations and with other nation governments and industry to find ways of accelerating the utilization of solar energy outside of the U.S.

Specific Commercialization Programs currently underway are:

1. Information accquisition, dissemination and exchange. Information about solar energy is being made available throughout the world through U.S. Embassy commercial offices; information exchange programs have been initiated between the U.S. and other nations.
2. Trade fairs, trade missions and seminars have been conducted in the industrialized countries and the developing countries to provide firsthand information about solar energy.

3. Needs assessments are being undertaken to assure that the solar technologies being proposed to developing countries are appropriate to the specific needs.
4. Joint ventures and other business arrangements are being encouraged and sought whereby accelerating the utilization of solar energy can be mutually beneficial.
5. Discussion of standards for hardware and installations are being initiated.
6. International meetings looking at both the technical and non-technical obstacles to the utilization of solar energy are being initiated.
7. Visitations by industrial and technical representatives from other countries to the U.S. have been underway and encouraged.

SERI ACTIVITIES IN INTERNATIONAL COMMERCIALIZATION

The following is a listing of some of the activities that SERI is undertaking as part of their work with the International Solar Commercialization Working Group (ISCWG) of the Office of Solar Applications of DOE.

- Assisting U.S. government agencies by providing an active centralized interface with scientific and industry associations.
- Creating an international commercialization bulletin.
- Operating as an industry broker to match foreign and domestic companies at the request of either side.
- Identifing industry standards and relating them to domestic technical standards.
- Developing a solar energy information data bank with a "user friendly" and efficient file management system which acquires, stores and makes the information easily accessable to industry users.
- Providing country profiles to facilitate market explorations.
- Producing a directory of key persons, companies and institutes in all countries.
- Collecting climatic data to facilitate market potential assessment.
- Working with the Department of Commerce in their trade fair, trade missions and seminars activities.
- Providing solar related reports and information to U.S. Embassy commercial offices and libraries around the world.
- Assessing technologies for their commercial readiness in international markets.

CONCLUSION

The above activities are part of an overall cooperative effort to encourage the wide scale use of solar energy. Such a cooperative effort today will hasten the day when solar will be considered the first choice for many energy applications and will accelerate the growth of markets to support the embryonic but growing solar industry inside and outside of the U.S.

UNESCO'S SOLAR ENERGY ACTIVITIES IN DEVELOPING COUNTRIES

T. Beresovski

Division of Technological Research and Higher Education, UNESCO,
Rue Miolis, 75700 Paris, France

SUMMARY

This paper describes the work that Unesco has been carrying out in the field of Solar Energy since the early 50's. Details are given of specific projects and meetings organised by Unesco and an assessment is made of this organisation's role in the development of this field of work.

KEYWORDS

Unesco, Conference, Congress, Seminar, Symposium, Mission, Feasibility Study.

The work of Unesco in the field of solar energy started in 1948 in connection with the study of energy problems in arid zones. As part of this programme, Unesco convened a conference in New Delhi in 1954 at which scientists from many countries presented analyses and plans for the use of solar energy in the developing world.

Unesco's role in the energy field was reemphasized in the early 1970's. During that time, a programme of educational activities, dissemination of information and promotion of international cooperation was launched, in which activities such as the following were included:

- the award of postgraduate fellowships for scientists from developing countries for research and study at advanced centres;

- support for the establishment of solar energy courses at the Centre Universitaire of Perpignan, France;

- technical assistance, under the Participation Programme, to a number of countries;

- the organisation and holding (with support to participants) of the Seminar on Solar Energy and its Applications in

Africa, Niamey, Niger, in October 1972.

These activities culminated in 1973 with Unesco's participation in the planning, organization and sponsorship of the International Congress "The Sun in the Service of Mankind", which took place at Unesco Headquarters from 2 to 6 July 1973. This Congress was preceded by an International Working Party of Solar Energy Experts, organized by Unesco and ehld in Paris on 27-29 June 1973.

Since 1973 Unesco has been involved with the promotion of solar energy research and development on a continuous basis, with increasing amounts of funds being allocated for the support of solar energy activities in developing countries. Solar activities were emphasized both through topical meetings and workshops of worldwide interest, and through focus on promotion of regional activities in Africa, Asia, Latin America and Europe.

To give examples of the topical gatherings of multiregional interest which were sponsored:

- Unesco, in cooperation with WMO, organized in Geneva, from 30 August to 3 September 1976, a Solar Energy Symposium which brought together meteorologists and solar energy scientists who, up till that time, were working only in their own narrow specialized field. Representatives from more than 64 nations and international institutions participated actively in the discussions and the preparation of the recommendations of this Symposium.

- Unesco, in cooperation with the North-East London Polytechnic, organized an International Solar Building Technology Conference in London from 25 to 29 July 1977. One of the main tasks of this conference was a critical evaluation of existing solar energy applications technology in building design and construction. The conference also dealt with broader issues of solar energy hardware under varying climatic conditions, with particular reference to developing countries which generally have fewer resources and less economic strength.

- Unesco sponsored a Workshop for practical training in the field of solar energy at Stuttgart University (F.R.G.) from 19-28 April 1977 and gave some assistance towards the travel costs of participants from developing countries. Similar assistance has been given to the Summer course on Solar Energy Problems organized at the Institut d'Etudes Scientifiques de Cargese (Corsica, 20 June - 2 July 1977).

- In October 1978, Unesco organized, in cooperation with Stuttgart University, a Workshop on technical and economical approaches to small-scale photovoltaic power utilization in rural areas. Stuttgart University has published a guidebook which shows the results of this Workshop, and which is particularly relevant to the needs of developing countries.

- In January 1979, Unesco sponsored a conference on Small Scale Energy for Developing Countries, which was organized by the Intermediate Technology Development Group and took place at the University of Reading, U.K.

- Unesco also provided financial assistance to the International Solar Energy Society for organizing the International Congress on Solar Energy which took place in Atlanta, Georgia, on May 20 to June 1, 1979.

With respect to regional activities, Unesco endeavours to cooperate with regional economic commissions and other International bodies as well as local governments to foster advancement of solar energy. It also uses for this purpose the staff of its regional offices for Science and Technology located in Jakarta (Indonesia), Montevideo (Uruguay), Nairobi (Kenya) with a branch in Dakar (Senegal), and in New Delhi (India).

Thus, within the scope of the regional efforts, the following projects were undertaken in Africa:

- In May 1975 Unesco participated in a UNDP mission to African countries to study the situation regarding solar energy activities in countries such as Senegal, Mauritania, Mali, Ivory Coast, Upper Volta, Niger, Chad and Cameroon. In addition, in 1977, Unesco organized an expert mission to Somalia to advise this Government on a long-term solar energy programme.

- Unesco has also assisted the Government of Rwanda in setting up a solar radiation measurements programme through support for the purchase of recording devices.

- The public solar energy company of Niger, ONERSOL, has been assisted by Unesco on a long-term basis and received help on the following projects:

  (a) Organisation of a solar energy course for African countries, which is to take place every two years.

  (b) Purchase of data collection instruments (solar radiation and temperature recorders).

  (c) Support for the travel of ONERSOL's personnel to Europe for educational visits.

- In 1977, Unesco provided a modest support for research activities in the field of solar energy to the Institute of Energy Research, Khartoum, Sudan.

- Recently, at the request of, and in cooperation with, the CEAO (Mali, Mauritania, Senegal, Guinea, Upper Volta, Ivory Coast), Unesco helped carry out a feasibility study to establish the parameters for the creation and operation of a Regional Solar Energy Centre expected to be located in Bamako, Mali. The Centre will undertake research on, development of, and training in solar energy, and will also be responsible for the production of solar devices (collectors, stills, dryers, cookers, solar pump assemblies). The proposed investment for the establishment and initial operation of the Centre is on the order of 40 million US dollars.

Similarly a number of activities were undertaken in Latin America. These include:

- Organization in November/December 1976 of a solar energy mission to Latin America. The mission visited Mexico, Costa Rica, Brazil, Bolivia, Chile and Argentina to collect information on ths status of solar energy research and development and to discuss with governmental officials and scientists possibilities for an accelerated introduction of solar energy utilization. A report has been published explaining the status of solar energy research and development in the countries visited, making recommendations and proposals, and listing the Institutions and scientists working in this field. This report has been disseminated by the Unesco Latin America Regional Office in Montevideo to all Latin American countries and all Resident Representatives of UNDP in Latin America.

- In February 1978, Unesco supported the second Latin American Solar Energy Congress in Paraiba, Brazil, by providing support to cover travel of participants from developing countries.

- Unesco also has been participating actively in the Inter-Agency Group Meetings on the development of alternative energy sources in Latin America. The last meeting of the Inter-Agency Group was held in March 1978. The aim of these meetings was to prepare a Study of Requirements and a Study of Capabilities to be used as terms of reference for future development and utilization of alternative energy sources in Latin America. These two studies were presented for discussion at a technical workshop of Latin American countries' energy policy-makers and energy specialists, and to other Agencies and Institutions interested in the problem. The Workshop, held at the end of 1978, provided guidelines for a large-scale renewable energy project for Latin America.

- In October 1979 Unesco helped organize a seminar on the conversion of biomass into fuel in Sao Paulo, Brazil. It was arranged with the cooperation of the Brazilian Government and the assistance of Unesco's Latin America Regional Office for Science and Technology. It recommended several follow-up activities which are under consideration.

Among the activities in Asia, the following may be noted:

- A group of experts was sent by Unesco to South and South-East Asia from 13 December 1973 to 5 January 1974 to undertake feasibility studies for the promotion of regional research programmes in that region. As a result, Unesco began negotiations with Member States in Asia regarding the establishment of a regional centre for energy, heat and mass transfer for Asia and the Pacific. This Centre was formally created at Bangalore (India), in February 1976, with considerable financial and moral support from Unesco. Currently the Centre is strengthening its activities. During 1976-1979 it organized workshops on heat and mass transfer, an annual post-graduate course on this subject, and published a Regional Journal on Energy, Heat and Mass Transfer.

- Unesco was one of the sponsors of the International Solar Energy Congress which was organized by the International Solar Energy Society in New Delhi (India) in January 1978, and participated in its preparation. Immediately after this Congress, an Asian Working Group on Solar Energy was organised by Unesco from 23-27 January 1978. Participants and observers from Asian countries as well as from ESCAP attended this Working Group. The purpose of the Working Group was to bring together specialists in solar energy to discuss regional activities and programmes, to identify the main directions of future development on a regional and sub-regional basis, and to make recommendations to coordinate programmes, so as to develop solar energy applications in the region. This was the first meeting of this kind dealing with solar energy held in the region, and led to the establishment of the Asian Network for Solar Energy.

- Unesco assisted in the establishment of a solar energy regional Information Centre for Asia and South East Asia at the Asian Institute of Technology (AIT), Bangkok, Thailand, by providing financial support to AIT for this activity.

Unesco has also been active in Europe both to foster development of solar energy in this region, and to help indirectly developing countries through strengthening of training opportunities and providing focal points for cooperation and advancement of technology:

- It cooperates and supports the International Center for Heat and Mass Transfer, Belgrade, Yugoslavia (ICHMT). This enabled ICHMT to sponsor seminars and summer schools with trained cadrees capable of helping development of solar energy.

- In October 1976, Unesco organized at the Genoa University a European Solar Energy Group. Invited participants were from many East and West European countries, as well as from several International Organizations. As a follow-up to the discussions of this Working Group, a final report was prepared, taking into account the status of solar energy development in Europe, reviewing the possibilities to organize European cooperation in solar energy activities, and including cooperation with other regions of the world.

- The above meeting was followed by the May 1978 Meeting in Varna, Bulgaria. This meeting was organized by the Bulgarian Academy of Science, with Unesco's assistance. As a result of this meeting a volunteer solar energy Executive

Working Committee was set up to work with Unesco and other international organizations to facilitate training and demonstration projects in countries where such activities are needed.

- In July 1978, Unesco organized in cooperation with the University of Perpignan (France) a Workshop on Solar Water Pumps. Unesco has also initiated the establishment of a solar energy Information Centre at the University of Perpignan by providing financial assistance for this purpose.

- In April 1978, Unesco participated in the Meeting organized by UNDP/UNEP in Geneva to discuss the findings of the UNDP/UNEP mission concerning the organization of a solar energy Regional Project of the Mediterranean countries. Unesco expects to participate actively in the preparation of this effort, aiming at a large-scale Regional Project.

Beyond the above activities Unesco also plays a wide role in disseminating technical information and assessing the status of various energy sources, including solar energy. Thus, in 1976, Dr. W. Palz wrote for Unesco a book designed to reach the broadly educated public, entitled "Solar Electricity, An Economic Approach to Solar Energy". This book was published in English, Spanish and has been translated into French. The Unesco's monthlies, "Unesco's Courrier" and "Unesco's Features" frequently carry articles on energy, and for instance the entire June 1978 issue of the Courrier was devoted to energy topics including solar energy; in addition, in October 1979, the entire issue of the Unesco magazine, "Impact of Science on Society", was devoted to assessing the world energy situation and the potential of various options, including discussion of solar energy in several of its articles. Also, the book "Perspectives on the use of different energy sources", by Prof. Z. Zaric, is in an advanced stage of preparation.

To further its assessment effort, in June 1979, Unesco sponsored the Second Information Forum on Fundamental World Energy Problems in Santiago, Spain. This Forum, in which participated many highly regarded experts, was devoted to a discussion of the perspectives of fusion and solar energy, with the main emphasis on their scientific and technological aspects. The purpose was to contribute to a better understanding of the potential roles of fusion and solar energy, for both developing and industrialized countries, for large and small-scale power generation, and for rural and urban development. The report of the Forum has been published and distributed.

Apart from the above-mentioned functions, Unesco provides financial support to Member States upon their request, in the form of short-term expert missions and staff missions to assist them in the creation and strengthening of their research and training institutions involved in solar energy activities. These missions also assist in the preparation of projects financed by extra-budgetary sources, such as UNDP or others. A number of fellowships have been and will continue to be provided to scientists from developing countries upon their countries' requests for studies of certain problems in the field of solar energy research and applications.

At present Unesco is cooperating with the United Nations system on the preparations for the 1981 UN Conference on New and Renewable

Sources of Energy to be held in Nairobi, Kenya. Specifically, it will contribute material to the Conference's Technical Panels on solar energy and biomass.

To conclude, it may be useful to present the recommendations Unesco has just received concerning the future direction of its effort. In accordance with the decisions of the 20th Session of the General Conference, a meeting of experts to advise on future Unesco programmes in the field of energy was convened from 3-7 December 1979 at Unesco Headquarters. This group of about 40 experts, serving in their private capacity, met during five days to advise Unesco on the direction and modalities of its future programmes in the field of energy. The group confirmed that Unesco's energy programmes in the past had played an important and even critical role in preparing scientists to deal with some aspects of the world energy situation and that Unesco should be expected to play an important role in the future.

Generally speaking, the current emphasis on small-scale sources of energy for rural and remote areas was commended, although it was emphasized that this should not be the exclusive direction of the programme. Importance should be given to the role of Unesco in promoting and understanding the social and cultural dimensions across the wider spectrum of energy problems. This inter-disciplinary approach is often lacking or inadequate in today's programmes and could be one of Unesco's major contributions.

The group confirmed that future Unesco programmes should include all energy sources but with special attention to the field of new and renewable resources and particularly solar energy. Unesco's experience and competence in education and training give the Organization a key role to play in these areas. Education and training programmes, including technician training, can be usefully expanded in all fields related to energy needs and problems. Unesco activities which are designed to promote and stimulate research related to the needs of developing countries have an important catalytic effect and should be continued and strengthened with the inclusion of state-of-the-art studies. It was pointed out that fundamental energy research should be strengthened both as an adjunct to education and in frontier areas of long range potential pay-off.

Special attention was paid to the role of Unesco in the field of information, in the light of the recent decisions of the Executive Board to launch a feasibility study in this area. It was brought out clearly that there is great need to strengthen information dissemination and communication, and that Unesco within the framework of UNISIST's programme can take useful initiative in this. Attention was given to the area of energy assessment and management in which a number of other UN agencies have programmes, and it was indicated that Unesco could play a useful role through support for training programmes.

Unesco's budgets for energy are very limited when considered in terms of the dimensions of the problems and the other major programmes (mainly national/bilateral) dealing with the existing problems. Because of this, the promotional role of Unesco will be stressed, involving programmes and activities in cooperation with other bodies, development of mechanisms on the national and regional levels (societies and networks) which can provide a multiplier effect, and identifying problem areas for research and development which can be followed-up by other more heavily funded programmes. On this level, Unesco, with irs non-biased approach and reputation for objectivity based on its scientific background has a special role to play.

POWER TO THE PEOPLE

B. Holdsworth

Holdsworth & Associates, Atelier St. Albans, 6 Spencer Street,
St. Albans AL3 5EG, Herts

There is a famous Woody Guthrie song that welcomed Roosevelt's New Deal. Hydro-electric power, coupled with oil and coal would create for America a peaceful, rich community working on ELECTRICITY.

Twenty years ago, I worked on the Gas Moderated Reactor programme and believed in this new Atomic Power that would transform the city slums of my childhood. Coal, our earthbound fuel, had been given, in name, to the people where pit-head baths helped give the miner the chance to use his kitchen differently. The miner would no longer be a mole, but a highly-paid engineer clawing the coal out with giant machines like I had seen in H.G. Wells' film "Things to Come". Power would be for the people.

CHEAP ELECTRICITY KILLED CHEAPER SOLAR/WIND POWER

Nearly two hundred years ago, when the first patent for a wave energy device was filed in Paris, the second industrial revolution started to move away from the direct use of water, in tidal mills, and waterwheels. Steam engines fired by coal and oil created immediately more power. The cheap oil of the 1940's and Woody Guthrie's clean white electricity stopped the sale of windmills and Solar hot water panels that had been a stable energy source in America's mid-West since the 1890's. By the late 1930's literally millions of wind generators and thousands of solar water heaters were rendered obsolete. Now overkill of cheap fossil fuel and world political power shifts has made some Americans turn back to the simple methods of self-generation. A wind power programme ranging both in size and type from 5 watts to 200 ft. machines with inbuilt real-time micro-processor controls to predict wind behaviour, and adjust accordingly, giving 2 megawatts of peak power for transmission into the electrical grid, is a sane example of balancing low scale (individual workshop) production with industrial manufacturing techniques using and developing existing skills.
Costs per installed kilowatt of these larger machines have fallen from £1,500 to £500 and it is reliably predicted to fall to £150 (using 1978 prices) by 1990 using mature industries using comparable manufacturing techniques.
Unfortunately the present British Government and its pet advisers are, in contrast, a bunch of Jeremiahs declaring that any Alternative Energy Source is a non-starter.

LIP SERVICE IS NOT ENOUGH

Lip service was paid by Government Scientist Professor Bondi seconded from the Arms Programme, to delegates at an International Solar Energy Society's meeting on December 4th 1979 but the real truth is that any crumbs off the Nuclear table will be little enough to go round.
It is important that those of us in the alternative energy field should be fully aware that however clever we are, here in the U.K. we are not being allowed to compete fairly with the mad dash for Nuclear Power.
Professor Bondi's own DOE figure of £155 million/year is at present being spent on Nuclear development compared to £4 million for the ten different alternative sources; and because there are so many different alternatives, spokesmen in Government and Industry who favour the NUKE lobby are giving as an excuse the problem of sharing cash. (ISES meeting 4.12.79).

It is claimed that by the year 2000 the U.K. will require 93 million metric tonnes of coal equivalent to bridge the energy gap between fossil fuel output and demand and that 20 fast Breeder Reactors will give us this power - (I wonder at what cost to us as a democracy) - the Nuclear programme it is claimed would create 660,000 jobs per year and cost £35 billion. Using the same source figures, the ten peaceful alternatives would create the same 93 m/m tonnes (of which Solar, wind, waves = 38 m/m tonnes) at a cost of £21 billion - with 1,520,000 jobs per year - and that's not counting a larger spin off of component manufacture.
One of the Jeremiahs was Mr. Glyn England, Chairman of the CEGB, who recently warned people against taking a romantic view of windpower - he informed a staff audience at their S.E. Region Headquarters that in order to get MAXIMUM output some 2,000 machines on the limited number of suitable hilltops in Britain would be needed to match the output of one power station. I was of the generation that was educated in the 3 'Rs' (reading, writing and arithmetic) and taking as an example America's NASA model Mod. 1 a 1500 megawatt station divided by 2 megawatts gives 750 sites, yet the wind power conch, now about to break wraps, will require but 20 sites.

COLDWATER ON WAVE POWER

The Central Electricity Generating Board acting as another vested interest for the Nuclear Power lobby is now pouring cold water on Wave Power. In 1977 and 1978, Ian Glendenning, leader of the CEGB Long Term Studies Projects report to the Labour Government of a study into wave power, said,

> "If wave power can be economically harnessed, its development potential along the U.K. west coast exceeds the present installed capacity of the CEGB - there are currently an average of 120 gegawatts of wave power being dissipated on the beaches - nearly 5 times the annual average demand on the CEGB".

Mr. Glendenning then went on to say that 20-30 km of sea could produce smooth power of 1500 megawatts - the equivalent of a standard power station.
Now in 1980, with the present Government's die cast onto the nuclear chessboard, the CEGB is more sceptical of being able to produce energy this way at a competitive price.

This obsession with price is further echoed in the Department of Energy Report No. 42 where it relegates wave energy as best to be regarded at present as a

possible insurance technology, saying that the future of wave power is entirely dependent on its competitive position in relation to other electricity sources which in turn, depends on the country's nuclear capacity. In a future where the use of nuclear power is unlimited, wave converters would have to compete economically with nuclear generating stations - wave generated electricity would cost 20 - 50p Kwh compared to 2p per Kwh from NUKE plants.
I ask at what cost to the safety of all the people in this nation!
I cast serious doubts on all this new found talk of pricing cheap by using nuclear energy.
The CEGB, strongly influenced by nuclear scientists, discounts its own appalling record in producing cheap electricity from Nuclear Power stations such as its own white elephant down at Dungeness.
However, back in 1978, Ian Glendenning's report did make the last Labour Government allocate money for wave development. By June 1978 a total of £5.4m was made, plus a mysterious grant for £300,000 to the International Energy Agency, to help develop a Japanese-controlled wave power experiment, based on an early British design.
Here was Britain, then grappling with the slide to an industrial slump but keeping inflation at 8% and having to close shipyards, aiding Japan - one of the most successful countries in the current economic struggle - to develop a British idea, and also build a research ship (the turbines and generator made by Centrax Ltd with British money).
Here we have an example of the politics of energy power being mismanaged, and many more examples are daily coming to light. At a time when successive Governments have let our Steel Industry decline - because it was primarily geared to automobile steel plate - there is a growing need for steel and pipes for combined Heat and Power installations, steel plate for wave machines, steel for windmills, steel for Solar furnaces - the alternative methods of energy production are a new industrial revolution in their own right.

DEMOCRATIC SOURCES OF ENERGY

The sun, wind, waves are all democratic sources of energy. They are freely given. They belong to all. Here we have a source of energy that gives heat to all regardless of race, colour, creed or sex and whether you live East or West of 0° meridian. We must ensure that no one fences in the sun.
But man has discovered another sun, one that is not democratic - a Nuclear Sun - we have brought the sun's heat to earth. It has to be fenced in. It is too hot to handle. Installations must be policed; its waste products are not dead, but can be simply changed into horror weapons of destructions and we transport the stuff on our rail and motorways. It is the new church. It is also the energy of dictatorship, of control by the few. If you criticize the Nuclear Programme then you commit blasphemy because you, in turn, question all sorts of assumptions and mythologies to do with our society; so you trust the scientist/ priest and insist that he knows best, that governments won't take risks with our lives; that they know best and we get that awful PASSIVE consumerism which leads to the microwave oven and microchip wristwatch. With the words "the silicon chip inside her head is switched to overload" The Boomtown Rats hit record "I don't like Mondays" had frightening overtones.
I referred to the Nuclear Lobby as the New Church - we who believe in Solar energy are of an older more humanistic religion. If we are to succeed it is essential that we develop educational programmes starting from primary school through all the grades to the education of local government planning officers and that we don't let Solar energy become a tool of Government but a tool for people to use. The social importance of Solar energy is that of all the alternatives, biomass, wind, wave, CHP, heat pumps, to name a few, Solar collectors (and indeed windmills) can be built in small craft workshops to suit the immediate low energy needs of the family. Here is small scale self-help that can return us to a more

decentralised human manner of living together in a community.
A prime example is China, where home produced cast pig iron production and heat from village pig manure silos gave China the chance to leap into the Industrial World whilst not having to serve its domestic market; and it is the whole question of scale that will become the battleground between those who want a centralist society and those who want a more open, interactive society.

BIG POWER = BIG BUSINESS

As a practising engineer, involved in endeavouring to conserve energy I also want to conserve people by keeping them employed while their standard of living rises. I want people to walk on both legs by using the best that can be derived;on the one foot, modern Science and Technology and, on the other, from tradional techniques.
With Solar utilisation the range of collection and storage equipment would be handled by all sections of society. Architects and engineers would construct buildings on passive gain principles; local authority building regulations should include such energy saving devices as solar panels, passive control, ventilated windows, absorption panels and heat storage as part of the foundations.
Larger undertakings would involve Industry and Government both local and national. There is no need for increasing Solar to the magnitude of Space Science proportions by beaming the sun's rays in from outerspace. Continual growth of the Gross National Product may well produce a saturation of material goods that will ultimately kill the spirit of man - but the development of all the other alternatives can give the earth a potential source of energy to enable future peoples to find the balance of living. At present our GNP is falling and the present British Government is trying to increase growth by putting more people out of work!

JOB CREATION VIA A SOLAR PROGRAMME

Solar utilisation not only creates jobs in the high energy and building fields, but can aid the development of new industries; in particular, I believe that in delta regions of Britain and abroad, fish farming can be expanded to large scale proportions by the use of solar covers.
To create development and activate people's involvement we must agitate for council byelaws and planning authorities to make it easy for everyman to deal with his own energy need. New enabling laws must be drafted that make energy saving as simple as taking a bath.
This country is one third the size of Texas with a population density 1½ times that of India; yet our laws related to environmental control have been spasmodic and sparce. It was not until 1960 that environmental pollution became a national issue. We have little legislation to control juggernauts, or control radioactive emmissions, or even passage of dangerous materials - yet I'm sure councils will find ancient byelaws and planning controls to stop Mr. & Mrs. Everyman putting up their own windmill or solar panel.

NEED FOR LOCAL AND NATIONAL GOVERNMENT HELP

Legislation from Central Government is too little too late; or is it too much too soon?.
For some eight years I worked within the District Heating Association to get a District Heating Enabling Bill through Parliament so that we could have waste heat pumped through utility lines to heat our homes. Wilson, Heath, Callaghan. They all listened and made polite noises. I thought that perhaps Margaret with here Science degree would at last understand the need. But with all the wealth of experience available in Russia, America, France, Denmark, Germany, Holland Sweden, Japan and Britain (in the steel town of Corby - the Jarrow of 1980 -

I designed a small district heating scheme for 2000 houses at a cost within the Government's yardstick) - Margaret Thatcher also has failed to act.
Now we have the Minister responsible for energy conservation, Mr. John Moore, saying that he has given 4 months of active consideration to the report on the Central Heat and Power Programme - with the remark -

> "We know CHP saves energy. We have yet to find out whether it saves energy economically, and to solve the wide range of technical, institutional and social problems which may arise from its development".

This statement is a disgrace - who is he kidding when the same Cabinet jumps into a programme of building nuclear plants that suffer from the China Syndrome and where the consequences of an accident are deliberately concealed from us?.
Furthermore, on Conservation Grants, the Department of Industry, with the DOE's sanction, is not renewing this grant when it comes up for review in June of this year . The original grant was for £25 million for 2 years from June 1978. Such sums, important to developing the Conservation Industry are miniscule compared to the billions of pounds being drafted in Nuclear Arms which can only lead to world destruction.
Sadly the greater the size and functions of the modern State, the less opportunity the average citizen has to take an important part in the conduct of its business.
I believe that the advancement of solar utilisation on the scale envisaged at this conference will take place outside Britain; this is not only unfortunate, but economic madness, that with nearly 2 million unemployed we are not creating a whole new group of industries. The People will not stay quiet for long and it may well be that the anti-nuclear, green peace movement will give the impetus we need to get Apollo's chariot rolling.

THE CONSERVATION OF PEOPLE

It is essential to conserve people. To do so I want a massive shift to decentralise industrial and urban organisations, with participation of individuals in management decisions both at work and in the running of our communities and our government. This cannot be with a move towards greater police involvement in our lives, or with civil servants becoming a secret Government that subverts Government Ministries and Parliament.
Professor Hall has told us how solar energy can help to create agricultural surpluses in a way that retains people on the land, rather than displacing them by importing over-mechanised systems.
It is technically possible to create a society that lives in balance with our planetory environment and conserves its resources. Resouce depletion, energy shortages, the "poverty gap", pollution, and population growth are soluble problems provided the human problems can be overcome.
I ask that we work in our respective fields to displace fear, to democratise our engineering and our science.

NATURAL SOLAR NOT UNNATURAL NUCLEAR

We cannot escape the physical inter-dependence of the World Community, and one of the planks of the International Solar Energy Society should be to propagate the view that solar energy is a form that can give energy independence for all nations.
Perhaps if I reach the age of 72 and someone suggest that I speak at the Solar 2001 Conference I might say-

"By the late 1970's early '80s it seemed the earth was dying. Nuclear accidents were creating social havoc and missile computers sensing high radiation were only just stopped in time. It was not easy for us to realise how mad the world had become. We tell our children of the senseless wars, like the 30-year Indochina conflict, of the limited struggles along the USSR-China border in the 1980's sparked off by the Afghan incident. Then the energy wars in Middle East, and the abortive push by the USA into Mexico.

It is hard now for us to understand that we suddenly came to our senses.
Perhaps it was the destruction of an American City by a faulty computerised nuclear missile in a practise run that shocked us all from our trance.

I am pleased to have been associated with the careful husbandry of renewable and extractive energy sources that put us all on a path of achieving a world peace worthy of human intelligence".

SOLAR ENERGY UTILIZATION AND ITS IMPLICATIONS FOR TRADE UNIONS

C. Levinson

International Federation of Chemicals, Energy and General Workers' Unions (ICEF), 58 rue de Moillebeau, P.O. Box 277, 1211 Geneva 19, Switzerland

The so-called "Energy Crisis" is not a crisis of availability, but a crisis of control. The reins of control in matters of energy policy have long been held by the syndicated interests of the multinational energy corporations and the financial nexus into which they are tied. Avaricious arabs or posturing politicians are not the real decision-makers.

Whilst these controls remain in the hands of those who stand to gain so heavily from shortage, energy will continue to be managed against glut. Oligopolistic control drives up energy prices unilaterally and so diverts capital from other investment sectors in an orchestrated disruption of economic development.

The same power concentration has been intensified in recent approaches to oil alternatives of all kinds. The domination of nuclear power by the capital-holding energy banks via the technology of the multinationals (such as Westinghouse) and by the monopolization of uranium supplies is a good example of this principle at work.

In a similar way, however, research of the softer technologies, and particularly of solar energy, has been stunted and diverted towards the development of centralised collection and generating systems. This route to solar power retains the primary requirement of such massive investment that only the controlling energy banks can fund it. By analogy with the past, the energy corporations will not be content until they have built a wall around the sun to meter out its energy at so many dollars per kilowatt/hour.

Trade unionists affiliated to the ICEF around the world are concerned with these matters both as consumers of high-priced energy, as defenders of democratic decision-making and as workers in the energy industries. In the latter capacity we object to the labour and expertise of energy workers being diverted from the social aim of providing widely available, low-cost power by the misdirection of policy choices to safeguard elite financial interests.

The trade unions, with all their faults, are undeniably the most democratically structured institutions of our time. Certainly, they are more democratic than the corporations and banks who control energy supply. They also bear a direct power relationship to the energy companies at the practical, operating level and through the ICEF are extending this relationship to the multinational context in which the companies move. This is a context beyond any other social or political apparatus-apart from the companies themselves.

Potentially, solar energy is one of the most democratically available power sources. It lends itself to dispersed, decentralised usage with relatively simple technology. The work of designing, manufacturing, installing and maintaining local or household solar collection systems would go a very long way to compensate for job losses in other sectors of industry, such as glass-making and auto manufacture. The same skills and equipment are relevant. Equally important, the nature of the industry is non-polluting and non-exhaustive.

Perversion of these desirable social ends by the drive to capitalise upward the first-fruits of solar research is already a highly visible threat. The trade unions in ICEF are determined to fight this tendency to the utmost and to extend democratic controls over energy futures from the workplace upward and outward.

INDEX OF CONTRIBUTORS

BERESOVSKI, T. 219-226.

BEVAN, E.G. 203-214.

CHEVALIER, Y. 197-202.

CHOULEUR, G. 93-106.

DASKALAKIS, D. 133-138.

FORREST, D. 63-68.

HALL, D. 5-20.

HEALEY, J. 23-62.

HOLDSWORTH, B. 227-232.

HUTCHINS, M.G. 79-92.

JAUHRI, S.M. 139-152, 153-156.

LEVINSON, C. 233-234

MICHAELIS, D. 107-114.

MURPHY, B. 23-62.

PAGE, J. 21-22.

RUSSELL-COWAN, A. 69-78.

SPEIDEL, K. 157-188.

SPONGBERG, R.C. 215-218.

STAMBOLIS, C. vii, 153-156.

STEWART, R. 23-62.

TABB, P.J. 123-132.

TABOR, H. 1-4, 189-196.

TUCKER, R. 115-122.

WATSON, S. 153-156.